T0224046

Textile Science and Clothing Technology

Series editor

Subramanian Senthilkannan Muthu, Bestseller, Hong Kong, Hong Kong

More information about this series at http://www.springer.com/series/13111

Subramanian Senthilkannan Muthu
Editor

Detox Fashion

Supply Chain

Springer

Editor
Subramanian Senthilkannan Muthu
International Chemical and Environmental
    Manager, Bestseller
Hong Kong
Hong Kong

ISSN 2197-9863   ISSN 2197-9871 (electronic)
Textile Science and Clothing Technology
ISBN 978-981-13-5227-0   ISBN 978-981-10-4777-0 (eBook)
DOI 10.1007/978-981-10-4777-0

Printed on acid-free paper

This Springer imprint is published by Springer Nature
The registered company is Springer Nature Singapore Pte Ltd.
The registered company address is: 152 Beach Road, #21-01/04 Gateway East, Singapore 189721, Singapore

This book is dedicated to:
The lotus feet of my beloved
Lord Pazhaniandavar
My beloved late Father
My beloved Mother
My beloved Wife Karpagam and
Daughters-Anu and Karthika
My beloved Brother
Last but not least
To everyone working in the global textile
supply chain to make it TOXIC FREE &
SUSTAINABLE

The original version of the book was revised: For detailed information please see Erratum. The erratum to the book is available at https://doi.org/10.1007/978-981-10-4777-0

Contents

Toxic Free Supply Chain for Textiles and Clothing

P. Senthil Kumar and S. Suganya

Abstract The existing structure of supply chain of textile industry is complex at every level that it leads for inter-dependencies across a network starting from raw material to manufacturing, clothes reaching customers. Complexity contributes to variability and uncertainty where a change in one element can have an effect on other elements. To feature such cumulative and combinatorial effect throughout the supply chain, good practice in labour standards leading to a legal minimal wage and realistic living wage, working hours, safety and integrated infrastructure have to be monitored by government bodies. The intention of this article is to serve idea on risk reduction measures for hazardous substances at every level through supply chain by identifying the toxic substances and its hazardous properties. For example, substantial shares of silver, triclosan, triclocarbon are released after the biocide treatment of textile from laundering. To avoid such human risk, the REACH registration is a source of limited knowledge on risk assessment of many substances used in textiles. Indian Garment Industry finds that inventory management, visibility, lead time, collaboration with private and government sector, technology as risk factors all over the supply chain. Based on target customer groups and scale size on production, most of the private companies are facing appropriate supply chain strategy for product offerings. The use of safer chemicals in the entire life cycle and production in apparel brings a substitution plan for hazardous chemicals. The aim of this chapter is to assure the use of safer chemicals and their substitution in acceptable range in the whole production procedures of textile products. That should meet customers demand for a cleaner production with no or less pollution in order to protect environment, new or modified environmental policies and regulations in order to protect workers and consumers health. Other regulative measures to aware customers by labeling toxic free clothes and suitable method for the pre-determination of toxicity in research laboratories must be taken into consideration by textile manufacturers.

P. Senthil Kumar (✉) · S. Suganya
Department of Chemical Engineering, SSN College of Engineering, Chennai 603110, India
e-mail: senthilchem8582@gmail.com

© Springer Nature Singapore Pte Ltd. 2017
S.S. Muthu (ed.), *Detox Fashion*, Textile Science and Clothing Technology,
DOI 10.1007/978-981-10-4777-0_1

Keywords Supply chain · Traceability · Logistic approach · Technology
up-gradation · Human risk assessment

1 Introduction

In today's world, the textile and garment industry having major contribution in
Gross Domestic Product (GDP) in economy. The recognized presence and high
stature in the global market build them strong supply chains to compete with global
trend. It gains advantage over world's trend and retailers by proffering the best
value to the customer demands. The apparel industry is having full of variability in
playing their role at every level of their supply chain. It considers refashion with lot
of structural, operational and performance differences all over the world. The more
interconnected textile industry is facing high exposure to shocks and disruptions.
The complex supply chain networks sporadically steer missteps and miscalculations
which have major consequences as their impact hinder the whole process right from
raw material selection to the customer desired product. In such case the supply
chain management (SCM) has become very critical to manage risk, dynamism, and
complexities of global sourcing. It also focuses on the instrumentation that information created by people and system generated. It is being controlled by flowing
out of sensors, actuators and radio frequency identification technology tags (RFID),
global positioning system (GPS) and more to detect their content and inventory
count itself [1].

"The supply chain will ultimately be measured based on its ability to produce
bottom-line results, such as EBIT and cost-to-serve techniques. However, with
significantly increased input costs, relying only on these measures can mask true
supply chain performance". A totally integrated supply chain is required for the
company to get gain the maximum benefits. It is assured by the interconnected of
customers, suppliers, IT system and its parts, monitoring objects. Extensive connectivity enables trending in apparel industry by introducing world class design
through world wide networks [2].

The aim of supply chain strategies and operation must be understood in order to
build the most effective considerations on cost containment, visibility, risk, customer intimacy and globalization. The focal point of supply chain management is
performance measurements that identify the success and failure in products reaches
customer. It deliberately knows the right sources to be selected, processed in the
global business environment. It is also responsible to satisfy or support corporate's
strategy. This approach consists of many organized entities which are highly
structured as well as the small scale, non-integrated spinning, weaving, finishing
and apparel-making enterprises and handicrafts having capital intensive and most of
the brand value, in the market by the handlooms and power looms [3]. It brings
contrary to the adversarial relationships, suggests seeking close relationships in the
long term with less number of partners and small scale handlooms and power
looms.

The structure of apparel industry performance can be enhanced qualitatively and quantitatively by offering more joint vendor programs, employment to small scale weavers, maintaining long terms relationships in terms to bear the rapidly changing market conditions and to present the best value in customer seeking products in the fastest way. This relationship aims for the new designs, technology up-gradation, sharing information and warehouse to emphasize SCM strategy. Such a partnership strategy regards as earnings before interest and taxes (EBIT) upgrades the analytical hierarchy process (AHP) to the beneficiary [4]. Although, top industries of this sector face infrastructural issues at every level which configures logistics related challenges. It can be managed under high pressure by the executives charged. The vulnerable supply chain finds it difficult in shortage of supply chain improvement projects. So expert suggests SCM to be transparent, demand-driven and more efficient.

Smarter supply chain implies set of responsibilities that executives can achieve in field that who are able to optimize complex networks of global capabilities. In such significant positions smarter system enables the limitation of human intervention by deciding automatically. It helps in an alternate evaluation against the complex and dynamic set of risks and constraints. Anyway supply chain framework has to be designed separately as per the requirements by particular companies to make it competitive in global market. The organized entities relieve players at every level having their individual brand value in terms to the total production of cloth and creating the largest employment [5]. This chapter discusses the structure of garment industry and its challenges with intensified product offerings, regulated strategies and framework of supply chain management with explored dimensions. It comprises of alignment, strategic depth, customer satisfaction with supply chain performance, supply chain network design, macro and micro agility, technology management, collaboration intensity and supply chain culture.

2 Structure of Supply Chain

The life of a garment from seed to sale is greatly varied from its origin to its final destination in the retail store. The end product is having hundred peoples hand in production. Here is the roadmap of toxic free supply chain from seed to store.

2.1 Energy

Any individual would inevitably find goods or cloths made in China, India, Bangladesh, Vietnam, Cambodia, etc. Supply chain in all textile processes has evolved to meet changing price and quality demands from the global marketplace. The long term goal of environmental sustainability is in account to the use of water, energy and natural resources into account the use of water, energy and natural resources [6]. Textile production is resource intensive. The production of textile-based goods requires a considerable amount of energy, from fiber production up to finishing. Energy is a natural resource, renewable or non-renewable, which can be converted into other forms. Prime renewable energy sources are hydropower, wind power and solar power, generally used for electricity production, as well as wood pellets and other plants which are used in combustion. Non-renewable energy resources are crude oils, gas, brown coal, etc., Nuclear power takes a special position as its by-product, the hazardous nuclear waste, requires tremendous time periods to degrade. Conversion from one form of energy to another is more or less efficient, depending on the energy value, the losses by transformation to energy forms (e.g. heat) which cannot be utilized in the system [7].

2.2   Stage I: Planting Raw Material

Fiber production is the basic stage in garment supply chain. The cultivation of cotton, silk, tenel, ramie, organic linen, milk silk, corn fiber, bamboo fibers, alpaca, soy silk, pineapple and recently banana for fibers, hemp are advisory for their high drought tolerant capacity. In addition, synthetic fibers such as polyster, nylon, viscose, acrylic and polypropylene recycled polyester, jute, black diamond fibre, polylactic acid fibre, lycra, lyocell, organic silk and organic wool etc., are also processed on the basis of waste to product technology [8]. A tiny cotton seed is the most used raw material as naturally coloured cotton that is accessible for colour variants like green, red, white and various shades of brown. An irrigation practice in country is responsible for cotton cultivation since it requires a great deal of water, a long frost-free period along with moderate rain water.

A number of ecological and sustainability of organic fiber requires heavy pesticide use. These issues affect the mass production, manufacturing, processing, packaging, labeling and distribution of organic textiles worldwide. As per Global Organic Textile Standard (GTS) framed the investigation policy over organic fibers about certification, labeling, licensing that ensure the organic status of textiles from harvesting raw materials through environmentally socially responsible manufacturing units [9]. It offers all the way to labeling provide credible assurance to the consumer. Issues in organic fiber due to lack of rainfall, high pesticide use, unstable climate, lack of financial facility, labour shortage, high wage rate, inadequate water supply, severity of diseases, lack of technology, high cost of inputs, low quality of fertilizers and pesticides and poor seeds. To predict the factors influencing cultivation of organic fiber Garret's ranking technique [10] has been used with the following formula:

$$Percent position = \frac{100(Rij - 0.5)}{Nj}$$

where,
   Rij = Rank given for the ith factor by the jth respondents
   Nj = Number of factors ranked by jth respondents
   This prediction score fulfill the scarcity of raw material by two ways.

(1) Biomimic to cultivate genetically engineered crops
(2) Synthetic fiber or man-made fiber

The example for genetically engineered is Bt cotton which produce Bacillus thuringiensis (Bt), a microbe that produces hundreds of proteins that are deadly to select species of cotton pests, but harmless to other organisms. It has some pros and cons. Synthetic fibers are perfect to replace natural fiber with low cost. Man-made fibers are made with names like acrylic, nylon, polyester and polypropylene. The success of synthetic fibers is due to low cost and inexpensive treatment methodology. Commonly used synthetic fibers are mass produced from petrochemicals to uniform strengths, lengths and colours, easily customized to specific applications. Another unique variety of fibers are blended fibers which are the blend of both natural and synthetic fibers. Anyways ecologically friendly options are recommended to incorporate plant fibers into fashion supply chain.

Significant strides have made toward managing supply risk by understanding part-material flows which fix 8–15% of their costs of goods sold [11]. Improved forecasting and planning extends the value chain through excess inventories. Systematic approach covers whether production materials are purchased directly for internal consumption and how much network materials spend or consumed by contract manufacturers. Demand Aggregation Programs [12] is associated with value drivers and sourcing strategies that differ based on market conditions, the targeted commodity groups and other factors. It is important to note that pricing is the standard form of benefits of leveraging.

2.3 Stage II: Yarning/Spinning

Each strands of fiber, shorter than the piece of yarn is twisted together into longer filaments to make yarn. The textile finishing procedure includes preparation and pretreatment, dyeing, printing and refinement of fabrics. The following steps are better established with chemical and non-chemical process. First, the natural form of raw textile materials has additives, dirt and other impurities, pesticides, worm killers etc. The removal these matters are called scouring done by adding suitable wetting agents, alkali and other chemical materials. It is done to achieve better wetting, penetration and dye uptake properties of fabric. The process of decolourization involves the removal of all natural colours in raw material to make surface orientation. The two forms of bleaching is oxidative bleaching using sodium hypochlorite, sodium chlorite or hydrogen peroxide and reductive bleaching using sodium hydrosulphite. After scouring and bleaching, Optical Brightening Agents

(OBA), available in different tints such as blue, violet and red, are applied to give the textile material a brilliant white look. Indeed, cleaning of bleach is must to neutralize the bleach.

2.4 Stage III: Fabric Production

The main stage of the supply chain is fabric production that transforms the yarn into fabric by weaving, knitting or a nonwoven process. The post processing of lengthwise kept yarn or twisted together is made either single or folded. To produce specific fabric, yarn is woven and then dyed. In knitting, yarn is inter-looped by latched and spring needles. It generates rolls of knitted fabric in particular apparel such as sweaters. Compression, looping, fixing, knotting, plaiting or twisting the yarn is achieved as non-woven processes and interlocking fibers by mechanical, thermal, chemical or fluid methods in order to produce the fabric [14].

2.5 Stage IV: Apparel Manufacturing

This stage consists of various processes including designing and stitching. Based on the market trends, customer needs and demand forecasting designs are to be made. Apparel companies either have their own designers or outsource from the various designer houses to preserve cultural values on fabric. In such way unique pattern pieces are created which is then used to cut the fabric in the specific shapes and sizes. The pieces are then assembled together into garments in the predefined manner as per the requirements of the design, through the stitching process.

Garment manufacturing companies also hire the local contract manufacturers for both the processes of cutting and stitching as higher executives instruct in every step [15].

Companies performing cutting in-house may economically upgrade the production process. Once the garment is prepared, labeled and shipped which are then allowed for the finishing process where it is cleaned, pressed. They are eco-friendly processed without chemical and pesticide followed by peroxide bleaching, green dyeing to improve sustainability of textile industry. The garment segment is the most labor-intensive and fragmented segment of the supply chain. Capital and knowledge requirements are being attractive for new entries. It is succeeded by innovative technologies that are integrated to attract global trend by 3D Printing Technology [16] for intuitive designs, less financial risk, less fashion effort and less production material. Processing and supply chain also concerns about long-term security by reducing CO_2 release into environment. A profitable choice for such production sharing usually falls on neglecting narrower product categories and frequently producing garments of both genders. The distribution to the respective retailer through appropriate logistics system and network should have higher proximity in global market.

2.6 Stage V: Retailing

The structure of retailing sector contains variety of retailing formats with structural varieties, distinguished approaches and operational differences. National/ International brands are sold in a variety of retail channels. Particular market segment seeking for aesthetic value holds major specialty stores accounted for 28.5% of all retail sales. Retailers and merchandise from a single company sell specialty garments and the large retailers sell wide range of assortments of garment products such as: Shoppers Stop, Globus and Westside [17]. Other apparel sales took place in discounters or mass merchandisers such as Wal-Mart, Kmart. In this category soft and hard goods are sold using an "everyday low prices" strategy. E-tailers and factory outlets are recently in demand selling goods through mail order. Another type is Full Line discount stores such as Big Bazaar and Vishal Mega Mart who offers low range branded products using the value based pricing strategy which attracts the majority of the population particularly the middle and lower middle class. Off-price stores are seasonal markets offering the designer products at much discounted prices [18]; however they are dumping assortments due to overstocking.

This chapter aims to review the current state of operations and recent trends across the fashion supply chain. It relies on the coordination between the flow of information and the flow of product and material. The information regarding the raw material needs to the suppliers is passed by three way gate system that heading backward from customers to the retailers and through them to manufacturers and to the suppliers in order to fulfill the customer orders, demands and market needs and trends. Companies performing synchronization between the information flow and material is important to make better profit.

3 Toxic Free Supply Chain for Textiles and Clothing

Fiber to fabric processing let us question, Are clothes toxic? It can be possible by use of toxic chemicals excessively in processing garment fibers and in manufacturing clothes. Synthetic fibers are safe is an illusion but full of invisible chemicals that the clothing industry prefers and customer rejects. The largest organ of human is primary concern as clothing clings. Industrial pollution, unorganized safety guidelines, chemical work place safeguard standards are ignored in most of the companies. During shipping, clothes are covered with formaldehyde to keep them free from wrinkling. Report discovering that China and Southeast Asia shipped clothes contains 900 times the recommended safe level of formaldehyde [19]. These synthetic materials are produced with toxic chemicals, and while they exhibit slow reaction, the long term accumulation. Then those accumulate cause air, water, and food contamination. It leads to the life threatening disease like cancer.

3.1 Astounding Chemicals in Apparel

Clothing that labeled wrinkle free may even contain perfluorinated chemical (PFC) used to make Teflon. It can cause numerous health ailments. Acrylic is another type of synthetic called for "wash-and-wear", "revolutionary time-saving leap" fabrics. "Wrinkle free" polyester fabrics are developed from xylene and ethylene. The mainstay of sportswear, swim suits, and thermal underwear is made from spandex and olefin. Olefin is produced by "cracking" petroleum molecules into propylene and ethylene gases [20].

Deadly chemicals used in textile factory include nonylphenol ethoxylate (NPE). It is well known in medicinal field as endocrine disruptors, like BPA. The use of NPE is restricted in Europe and America for killing a lot of fish and wild life when dumped into waterways. This negative effect on sea wildlife receives intercontinental attention by textile executives. But textile factories from China and Southeast Asia are free from such restrictions. Toxins like formaldehyde, brominated flame retardants, and Teflon are in practice to provide "non-iron" and "non-wrinkle" qualities. Insecticides are even applied in the name of good health. Anyway today's clothing is manufactured using an astounding 8000 synthetic chemicals. This is the main reason for the interaction between skin and chemicals creating problems like infertility, respiratory diseases, contact dermatitis, and cancer [21].

Additionally, textile industries use volatile organic compounds (VOCs) that are ubiquitous in the environment. They contribute to CO_2 emission, poor indoor air quality and carcinogens. U.S. Environmental Protection Agency estimates concentrations of VOC's can be as much as 1000 times greater indoors than out. Such evaporating chemicals from many fabrics as cleaning agents can cause chronic and acute health effects. Dioxin-producing bleach used by textile industries is an active ingredient for Agent Orange. Lastly black clothing and dyes for leathers often contain p-Phenylenediamine (PPD), which can produce allergic reactions. In bedding, nightwear may accumulate carcinogenic flame retardants [22].

The following list of chemicals used in textile production are benzenes and benzidines; methylene chloride, tetrachloroethylene, toluene and pentachlorophenol. However VOC's are also found in drinking water. This is not a good indication because it can enter to ground water from a variety of sources as like from textile effluents to oil spills. Maximum contaminant level (MCL) has been established for each chemical or additives. Easy care (Formaldehyde), water repellent (Fluoropolymers), Flame Retardants, Bacterial and fungicidal chemicals (Triclosan and nano particles) are reported to cause skin/lung irritation and contact dermatitis. Other scary toxins include sulfuric acid, urea resin, sulfonamides, halogens, and sodium hydroxide [23].

Table 1 The list of common chemicals and its application

S. No.	Name	Chemical formula	Application
1	Potassium permanganate	$KMnO_4$	Helps for color out from garment
2	Caustic soda	$NaOH$	For bleaching without changing colour of the garments
3	Soda ash	Na_2CO_3	Act as washing soda and softening water It creates alkaline medium for the breakdown of pigment dye
4	Acetic acid	CH_3COOH	To neutralize the garment from alkaline condition
5	Sodium meta bisulphate	$Na_2S_2O_5$	Act as a reducing agent
6	Pumice stone	–	Act as an abradant in washing cycle
7	Enzymes: Textile enzyme N1000 Neutral cellulose Bio polish enzyme Cellulose enzyme	–	Develop "Bio-polishing" effect, "Anti-pilling" properties, increases the color fastness and rubbing fastness properties
8	Bleaching powder	$Ca(ClO)_2$	Helps white clothes to retain their original color time and again
9	Hydrogen peroxide	H_2O_2	Antibacterial and antiviral properties and kills mold, mildew
10	Silicone softener	–	Reduce static cling, soften laundry and makes ironing easier
11	Anti staining agent	–	Desizing and washing in denim rinsing.
12	Sodium bicarbonate	$NaHCO_3$	Act as anti-pollution agents
13	Cationic/nonionic flax softener Ethers, polyglycol esters and oxiethylates	–	Creating softer handle over the garments, enhancing performance of the surfactant used in the bleach bath
14	Salt: Rock salt	–	To remove impurities from the Garment fabric surfaces, act as an electrolyte for migration, adsorption and fixation of the dyestuff to the cellulose material
15	Buffer: Monosodium phosphate	NaH_2PO_4	For bleaching and dyeing fabrics
16	Stabilizer: Sodium silicate	Na_2O_3Si	Act as a fixing agent, enhancing performance of the surfactant used in the bleach bath
17	Optical brightener: Resin	–	Improve the brightness of garments

Anyway, urgent action must be taken to replace hazardous chemicals with safer alternatives in clothing (Table 1).

3.2 Impacts of Toxic Chemicals

The apparel industry has been condemned as being one of the world's worst offenders in terms of pollution. Not only synthetic fibers, natural fibers have their problems, too. Organic farming of natural resources is pesticide intensive. For instance cotton is the most pesticide intensive crop in the world. It occupies up a large proportion of agricultural land which is devoured by local people. In that way, it kills or injures many of them.

Chemical defoliants (Tribufos, chemical name, S,S,S-tributylphosphoro-trithioate) sprayed on plants to cause the leaves falling off. It is an emulsifiable concentrate (EC) cotton defoliant mix with the insecticide to accelerate the aging of cotton leaves and to prevent losses from boll rot organisms [24]. Reports are showing that the exposure of tribufos and its metabolites in ground water let them bind to the soil and appear to be immobile. But it can potentially contaminate surface water. The high soil/water partitioning of tribufos indicates that runoff will generally occur via adsorption eroding soil as opposed to dissolution in runoff water.

The toll on both the environment and human health is herbicides which are sometimes used to aid mechanical cotton harvesting. Those chemicals retain in clothes even after finishing which can disturb the lifetime of the garments. Technology is able to produce natural abundance artificially. Likely, genetically modified cotton crops, resistance to insects, herbicide and disease add environmental problems at another level. The cotton industry is still largely dependent on cheap labour. Despite organic harvesting, the raw cotton is dyed; resulting fabric creates toxins that flow into our ecosystem. Finally it is to be sown into a T-shirt. From wastewater emissions to air pollution and energy consumption, the textile industry weighs heavily on the environment.

Fiber to fabric production is successful by bleaching, dyeing, and finishing. Both agricultural and textile workers are exposed to organophosphate sheep dip problem. It uses yet more energy and water, and causes yet more wool pollution. Synthetic chemicals made from petrochemicals are non-biodegradable, and so they are inherently unsustainable [25]. It also liberates green house gas (nitrous oxide) 310 times more potent than carbon dioxide. Lubricants used in clothes needs large amounts of water become a source of contamination. Lately, artificial fiber made from wood pulp seems more sustainable. For this purpose, the more planted tree is eucalyptus, which draws up remarkable amounts of water, causing problems in sensitive regions. Caustic soda and sulphuric acid treated wood pulp is detrimental to the environment on many levels. Chemical residuals that evaporate into the air we breathe or are absorbed through skin. Some of them are carcinogenic, trigger allergic reactions in people. All the processing and dyeing units need in garment industry must have an ETP (Effluent Treatment Plant) to avoid polluting the water and the soil of our planet.

3.3 Problems with Synthetics

- Skin—the largest organ absorbs toxins and they bypass liver which is responsible for removing toxins.
- Petrochemical fibers restrict and suffocate skin that shutting down toxic release.
- "Tipping point"—Total toxic burden induced in body for triggering the onset of disease.
- It is left with two factors. One is toxic buildup in body. Second is interaction between multiple chemicals to create even worse problems than the individual chemicals.
- Skin rashes, nausea, fatigue, burning, itching, headaches, and difficulty breathing are all associated with chemical sensitivity.
- In the tag of anti-cling, anti-static, anti-shrink, waterproof Perspiration and mildew resistant, Chorine resistant linked to a 30% increase in lung cancer.
- Flame retardants can cause thyroid problems, brain damage, ADHD symptoms, and fertility problems.
- Permethrin—insecticide used in military uniforms failed to prove safety.

4 Suggestions by Researchers

The effectiveness of defoliants (sodium chlorate, ginstar, starfire, dropp, def) and their combinations are studied by researchers. The report showing that there is no significant difference in the untreated and defoliation treated. But it varies from region to region even within fields, years. The report on the performance of growth regulators are inadequate yet optimal application guidelines is drawn between existing products as well as new products such as PGR IV, Glyphosate, and Atomik [26]. However, field study has noted that consistent and significant increases in cotton yields. The following table is drawn with the distinct sources, literatures and field survey (Table 2).

This suggestion is made from manufacturers label based on the application rate, intervals to harvest and tolerance established by the Environmental Protection Agency for these chemical residues [27].

4.1 Sustainability in Production—an Alternate Source for Toxic Free Supply Chain

Sustainability defines meeting human needs without overwhelming nature or society. It also prevents user from health effect due to toxic agents. To improve sustainability organic textile covers the cultivation of raw material, mass production, manufacturing, processing, packaging, labeling and distribution of organic textiles worldwide.

Table 2 Significant chemicals and its application on organic field

Name	Application
Defoliants	
Sodium chlorates (with fire suppressant)	a. Apply 7 days prior to harvest b. Do not add insecticides or other chemicals unless specified on label c. Do not graze treated areas or feed gin trash to livestock
Sodium cacodylate	a. Add 1/3–2/3 pt of nonionic surfactant b. Apply 7–10 days prior to harvest
S, SS (Tributyo)	a. Requires 5–7 days minimum for leaf drop (10–14 days may be required)
Tributyl phosphorotrihiolite	a. Higher rates may be necessary b. Certain phosphate insecticides can be mixed with Def or Folex for late-season boll weevil control
Dimethipin	a. Use with crop oil concentrate 5–7 days later b. Avoid spray drift
Thidiazuraon	a. Apply when top harvestable boll is mature at least 5 days prior to harvest
Desiccants	
Paraquat	a. Apply when >80% bolls are mature. Use a nonionic surfactant 7–10 days before harvest b. Do not pasture livestock in treated fields within 15 days after spraying. Livestock should be removed from treated areas 30 days before slaughter
Endothall and paraquat	a. Apply 3 + days before harvest b. Use nonionic surfactant at 1 pt/gal mix c. Recommended for stripper harvested cotton at high rates for considerable growth
Growth regulators	
Mepiquat	a. Apply when plants are in the early bloom stage and about 24″ tall b. Up to 4 days past early bloom, low-rate applications are optional, allowing discontinuing the applications (beginning at pinhead square) if stresses occur c. Pix should not be applied if plants are under severe stress (from weather; mite, insect or nematode damage; diseases and/or herbicide injury) d. If drought stress occurs when Pix is applied, or after a full rate (0.5–1 pt) has been applied, results will be less than optimal
Ethephon	a. Apply when 50–60% or more bolls are open (that is, when the top harvestable boll is mature) b. Prepare crop for defoliation, treat with Prep 4–14(+) days before defoliation treatment c. Prep can be mixed with Def, Folex, Ginstar, Harvade, methyl parathion, guthion and malathion d. Prep and Ginstar has an effective tank mix in some cases e. Temperatures 65 °K or lower after treatment will delay boll opening

4.1.1 Sustainable Supply Chain

(1) Critical resources are conserved which generates business value by unlocking opportunities.
(2) Firms with sustainability practices get rewarded by capital market shareholders and stakeholders.
(3) Impending climate legislations let us understand the importance of supply chain risks.
(4) Sustainability helps in building supply chain to avoid environmental impact and risks on water usage and carbon.
(5) New products having carbon/sustainability label will increase in value with consumers getting more knowledgeable.
(6) Sustainable supply chain is a great initiative by all major textile brands implementing measures within their own facilities.
(7) Sustainable supply chain exists where the majority of the environmental footprint exists.

The Zero Discharge of Hazardous Chemicals (ZDHC) is an assessment specific to the apparel industry. The goal is not only eliminating hazardous chemicals but improving sustainability throughout the supply chain by prioritizing the substances. Sustainability becomes mainstream focus across the entire supply chain that makes supplier choices. It also qualifies suppliers to report and demonstrate sustainability measures. Vertically integrated firms as early adopters of sustainability use market advantage through easy visibility throughout the supply chain.

Chemical free clothes are welcome having no access to clean drinking water, water scarcity and water pollution. Washing of textiles releases alkyphenols and perflorinated chemicals (hormone disruptors) have been found in animals, birds, whales to polar bears all over the globe. So, manufacturers create safer supply chain with organic textiles. This is for public knowledge that organic cotton as raw material is going to remain a very small percentage. It is expected that the emphasis shift toward other sustainable materials. Emphasis extends greatly on the emphasis shift toward other sustainable materials [28].

5 Role of International Standard for Organization (ISO) in Toxic Free Supply Chain Management

Regulations are adopted to ensure that citizens, businesses, and public authorities can readily identify their rights and obligations. The general obligations are

- To state the full fiber composition of textile products;
- Minimum technical requirements for applications for a new fiber name;
- Requirement to indicate the presence of non-textile parts of animal origin;

- Exemption applicable to customised products made by self-employed tailors;
- Reporting on the implementation, review clause, and study on hazardous substances to be undertaken by the Commission.

Each country has its own regulatory agencies in charge of developing policies, development, export promotion, and trade regulation in respect to the apparel and textiles industry. Market available products must be labeled. The indication of the fiber composition of a product at all stages of the industrial processing and commercial distribution is recommended. It scrutinizes the product quality and stay away from fake products. The Regulation does not cover size, country of origin, or wash/care labelling.

The standard for quality management is certified by International Standard for Organization (ISO). It is been administered by the accreditation and certification bodies. It clarifies what an organization needs to fulfill to meet customer and regulatory requirements, quality management system, and provides guidelines for performance improvement. More than 150 countries joined together in implementing renowned international standards which covers a set of procedures to monitor the process and its effectiveness [29]. To keep pace with the market scenario, industries are in need to rely more on innovative fabrics. Innovations in the fields of nanotextiles, nonwovens, electro textiles, medical textiles, and geo textiles are providing new opportunities for the manufacturers and tap the market. Simultaneously, these opportunities also encompass them with critical challenges that few of the chemical residues from fiber preparation sometimes emit pollutants during heat setting processes, dyeing operations and treatment plant operations. It may sometimes lead to volatilization of aqueous chemical emulsions during curing stages [30]. Research organizations today engage in many activities to resolve these issues. They have to be engaged by industrialists.

5.1 Regulations for Importing/Exporting Textiles

All textile industries have to fulfill the forms Consumer Product Safety Commission (CPSC) requirements for flammability. Wearing apparel must have labels specifying content and instructions for care. All apparels must have either labels indicating the country of origin, if this is not feasible, the fiber content (yarn, thread, wool) or packaging in such a way that country of origin is discernable to the ultimate purchaser.

Labeling or determining country of origin for mixed products is complicated. In such case other information is required. The purpose is

1. To protect consumers by providing accurate information on product origin, construction, quality, and care;
2. To protect businesses from unfair competition deriving from false representations of product content, origin and quality.

Every country must be a member of the WTO. Imports must be provided national treatment as per article 3 of the General Agreement on Tariffs and Trade (GATT). Technical Barriers to Trade (TBT) is the WTO Agreement, prohibits technical regulations (including labeling requirements) that create unnecessary obstacles to international trade [31]. That can fulfill certain legitimate objectives, which include national security, prevention of deceptive practices or protection of human health and safety. Additionally it cannot be unfair to domestic producers.

5.2 The Global Standards

The Global Organic Textile Standard (GOTS) is the worldwide leading textile processing standard for organic fibers, including ecological and social criteria, backed up by independent certification of the entire textile supply chain. The aim of the standard is to define world-wide recognized requirements that ensure organic status of textiles, from harvesting of the raw materials, through environmentally and socially responsible manufacturing up. Labeling is used in order to provide a credible assurance to the end consumer. Textile processors and manufacturers are enabled to export their organic fabrics and garments with one certification accepted in all major markets. The standard covers the processing, manufacturing, packaging, labeling, trading and distribution of all textiles made from at least 70% certified organic natural fibers. The standard does not set criteria for leather products.

Organic certification of fibers must be done on the basis of recognized international or national standards (IFOAM family of standards, EEC 834/2007, USDA NOP). Global Organic Textile Standard (GOTS) provides certification of fibers. A textile product carrying the GOTS label grade 'organic' indicates a minimum of 95% certified organic fibers presence in it. Whereas a product with the label grade 'made with organic' indicates a minimum of 70% certified organic fibers [32].

At all stages through the processing organic fiber products must be separated from conventional fiber products. All chemical inputs (e.g. dyes, auxiliaries and process chemicals) must be evaluated and meeting basic requirements on toxicity and biodegradability/eliminability Prohibition of critical inputs such as toxic heavy metals, formaldehyde, aromatic solvents, functional nano particles, genetically modified organisms (GMO) and their enzymes The use of synthetic sizing agents is restricted; knitting and weaving oils must not contain heavy metals Bleaches must be based on oxygen (no chlorine bleaching). Azo dyes that release carcinogenic amine compounds are prohibited.

Discharge printing methods using aromatic solvents and plastisol printing methods using phthalates and poly vinyl chloride (PVC) are prohibited. All operators must have an environmental policy including target goals and procedures to minimize waste and discharges. Wet processing units must keep full records of the use of chemicals, energy, water consumption and waste water treatment, including the disposal of sludge. The waste water from all wet processing units must be treated in a functional waste water treatment plant. Packaging material must not contain PVC. Paper or cardboard used in packaging material, hang tags; swing tags etc. must be recycled [33]. Raw materials, intermediates, final textile products as well as accessories must meet stringent limits regarding unwanted residues.

5.3 Certification of the Entire Textile Supply Chain

a. Fiber producers (farmers) must be certified according to a recognized international or national organic farming standard that is accepted in the country where the final product will be sold.
b. Certifiers of fiber producers must be internationally recognized through ISO 65/17065, NOP and/or IFOAM accreditation. Operators from post-harvest handling up to garment making and traders have to undergo an onsite annual inspection cycle and must hold a valid GOTS scope certificate applicable for the production/trade of the textiles to be certified.
c. Certifiers of processors, manufacturers and traders must be internationally accredited according to ISO 65/17065 and must hold a 'GOTS accreditation' in accordance with the rules as defined in the 'Approval Procedure and Requirements for Certification Bodies'.

6 Restrictions at Supply Chain

The following difficulties are being faced in supply chain in textile industry for yarn, cloth, apparel, garment, industrial yarn, etc.

(a) Difficulties occur largely at larger the distances in reaching the materials at proper time at customers end.
(b) Larger the distance, influence in transportation cost: It is unable for some customers to get the right raw material from the right resource because of high cost of transportation.
(c) Improper production planning becomes more erratic at both manufacturer and consumer end.
(d) It is more likely uncertain when there are fluctuations in demand of consumer product.
(e) The staggering government policies become problematic in taxation.
(f) In case of raise in market demand, the supply becomes more critical because of non-availability of trucks, manpower and resource problem. At that time, the manufacturers are unable to cope up with the growing demand because of their limited capacity.

(g) A thorough vision in planning is must to maintain demand supply.

(h) Outsourcings are being done to meet the demand supply through proper supply chain.

(i) During the market demand, the transporters, the concerned loaders and unloaders start demanding more wages disturbing the chain link.

(j) Increasing the uncertainty in the international market, the customers start stocking of the materials and hence, subsequent problems arise in logistics and distribution.

(k) Routine distribution is necessary to manage limited supply. Regular follow up of a customer profile helps to achieve routine distribution.

(l) Importances are being empathised on valuable customers for upkeeping the customer's business online.

(m) In case of special product requirement at remote place where logistic becomes difficult but to fulfil the customer need it requires to know the presence of other customers in the nearby areas, so that proper distribution can be made at a reasonable logistic cost.

(n) During off-season, weather changes, bad road condition, natural calamities, etc., it becomes difficult to dispatch the material at customers end in time.

(o) By keeping the proper information with the dealers and the customers, it is easy to avail adequate materials and dispatching the same.

6.1 History of Revolution in Textile Trade

Due to globalization, the competition of textile industries in terms to supply their goods and services were way behind in the quality. Intensive importing of goods hugely affect by control on production, licensing restrictions along with high protective policies, which had fostered monopolistic trends within textile industries. The state, national and global regulations are promoted for inward looking and the disappearance of trade barriers. Trade even impacts in textile industry, outlines the trends in trade, issues facing in clothing trade in textiles. Some of the policy measures are need to be considered to overcome this.

Newly industrializing countries (NICs) are Japan, Hong Kong, South Korea and Taiwan added other benefits including improved quality level, flexibility of production and stylish merchandise to offer products at lower price. The imposition of quotas by well developed nations made NICs to shift their production to other less developed countries. Trade Development Council survey helps to improve the quality of goods and exporting polices that has to be revised. This segment attempts to delineate the changing patterns of trading scenarios.

The arbitrary quantitative export restriction was first imposed in 1950s in the Asian countries. Short Term Arrangement (STA) is the first wider protectionism measure taken in 1961. Followed by Long Term Arrangement (LTA) was introduced for severe trade restriction and employment in western textile industries. The regime of the Multi Fiber Arrangement (MFA) was started all over the world effectively in 1974 [34]. It provides guidelines for the member nations. After that, the Uruguay round (for trade talks) had announced for bilateral agreements that few of the developing countries can't follow. But in later, it culminated the formation of World Trade Organization (WTO) in 1995. Import quantities of textile products were governed by the Agreement in Textile and Clothing (ATC) that provides eventual elimination of quotas. This in fact is reason for liberalization in trade activities with the anticipated application of free trade norms.

International textile business is claiming unemployment, hardship with resulting bankruptcies. It is threatening both importers and exporters of both developed and developing nations. WTO is a positive force sorts out issues in apparels. The liberalization of clothing and textiles has been controversial in both developed and developing countries. But still, the removal of tariff and non-tariff barriers and to accelerate the process of globalization is achieved by WTO. The dual measures of rationalizing and restructuring production must be confined in order to integrate regional and global networks of production. It will give aggressive result on export.

6.2 Prediction on Toxicity

6.2.1 Hazard Assessment

Hazard assessment is a segment in which toxicity and dose response, exposure to water and air is pre determined. Dermal absorption is performed with the help of animal studies. The various concentrations ($\mu g/cm^2$) with respect to time (hr) are performed by considering significant skin residue before and or after washing with soap. Mean total dose recovery through urine, fees, dermal dome and duodenum; mean absorbed dose is finally analyzed. Next, dose response assessment is experimented to check the special sensitivity to infants and children by safety factor. Toxicity Endpoint Selection is a key factor which conducts the previous risk assessment.

Exposure scenario also includes the risk assessment on inhalation. Dietary exposure includes food and drinking water. Risk characterization in food considers plant/animal metabolism to depict the qualitative nature of the residue in them. Tolerance reassessment has been conducted by storage stability (Storage intervals vs conditions of samples of cottonseed).

Health Advisory Level of astounding chemicals [35] has been established to avoid the discharge industrial processed water to the major water body. Pesticides in drinking water show unacceptable aggregate risk in all food exposures and other non-occupational exposures. To estimate the concentrations of toxic chemicals, groundwater monitoring system has been introduced.

7 The Consumer Adoption Behavior to Eco/Sustainable Fashion

Sustainable fashion concept is a central theme that investigates about how much customers are aware of their wardrobe. Its constitute dimensions are considerable challenge varies with demographics. Both lifestyle and demographic variables influence in perceived awareness deficiency and that provides a plausible explanation to the attitude-behavior discrepancy. Corporate Social Responsibility (CSR) is a key factor which has natural responsibility. Determining their activities and role in customer's decision is superficial. It clarifies about the following criteria includes organic resources, designing, artisan, recycle, custom and fair trade. Despite environmental sensitivity, sustainable fashion in the market place introduces risk associated with the product availability and heterogeneous findings on adoption influencing factors. A middle range theory [36] is developed as a result of the investigation between barriers and enablers of consumer adoption. It suggests that thwart adoption and engender adoption is based on the consumers disposition and offering characteristic; marketers competencies and consumers benefit. Therefore, sustainability-oriented marketers should avoid green-gouging, ensure product availability, clearly articulate adoption benefits, and develop offerings that are functionally and environmentally sound.

Fast fashion is one of the driving forces that persuade consumers buying more clothes regardless of their need. This concept is contradicting to sustainability and also adopts mass production and consumption. On other hand, excess consumption leads to the disposal of superfluous clothing. Clothes made from organic cotton involving recycle initiatives are environmentally friendly business strategy. Yet many consumers are reluctant to adopt sustainable changes in their clothes. Sustainable practices in apparel industry must be investigated in a particular interval. It should facilitate socially responsible consumption and to explore consumers perception on sustainable fashion.

Eco friendly clothing (EFC) contributes the following factors such as fashion orientation, shopping orientation, environmental concern, and eco-friendly behavior. These are independent variables. Purchase intention regarding EFC is a dependent variable. A series of linear regression analyses indicates that one fashion orientation factor, two shopping orientation factors, and three environmental concern and eco-friendly behavior factors are significantly related to consumers' purchase intention. From this analysis, executives suggest that effective marketing approaches as well as strategies for developing successful sustainable apparel products must be paid attention by consumers. Shoppers are more likely motivated to adopt EFC that align with their financial and sustainability interests. Labels like 100% cotton, natural, sustainable or environmentally friendly attract shoppers in different ages [37].

Although consumers express more concern about increasing prices at retail than about environmental and social issues. Only 51% of them are "green consumers" and the rest is confused about marketing terms. It is retailer's responsibility of bringing consumers' willingness to trust eco-friendly claims, and to minimize environmental footprint continue to represent areas of challenge and opportunity. The fiber content label, garment hang-tag and the packaging, made in label are most likely to influence purchase decisions. The current economic condition makes buyers less likely to afford EFC but willing to buy locally made products. A strong connection between the economic situation and decision has gained momentum as a national movement.

7.1 Dual Views on Sustainability

Sustainable refers to the durability or longevity of apparel. The environmental tags, "does not harm the environment" and "made from renewable resources" which are associated with the word "sustainable". Life span of cloth is a main perception looking forward by consumers that define the apparel quality. Disposable inexpensive clothing has a possible diminishing attraction. The long-term financial costs of fast fashion have coupled with potential environmental impact that expands sustainable cloth offering. Embracing sustainability means to reduce the environmental footprint of apparel aided by clear financial savings.

Go-green revolution from food to consumer products must adopt the values of sustainability, transparency, and authenticity. Yet, does the next generation of buyers across all demographics demand sustainability and transparency in terms to clothing?

Lack of information about apparel supply chains and transparency on large scale doubt the above mentioned question. Amazon's dream team says that most people are in "pre-awareness phase" when it comes to ethical fashion. Social norms and social influence are critical for really transforming consumer awareness and actual behavior change.

The Prius effect—recognition as a person who is thoughtful about environmental or energy issues. Clothing is indicating signals and creates the potential for status and for the social influence effect. Buy less and reuse and repair more is a mantra for consumers. Conscious Collection has a distinct green tag that communicates explicitly consumers. Fashion Revolution focus on empowering consumers to talk about brands and making sustainability cool [38].

7.2 Advices to Consumers for Detoxing Wardrobe

a. Switch to natural detergents that take a few washings to remove the residual toxic detergent ingredients completely.
b. Clothing labels are the indicators to avoid synthetic materials such as Rayon, Nylon, Polyester, Acrylic, Acetate or Triacetate.

c. Customers can avoid wrinkle free and preshrunk items.
d. Specialty stores help customers to stay in the 100% pure cotton or hemp clothing.
e. Natural detergents can replace the purpose of using baking soda to help neutralize new clothing chemicals.
f. Perchloroethylene (PERC) is used in dry cleaning. That can be avoided.
g. Wash and dry clothes purchased from thrift stores because they may be sprayed with some chemical before they are put up for sale.
h. Silver nanoparticles in name-brand clothing is appreciable that can easily absorb toxins due to their miniscule size.

8 Goals to Achieve in Future

The minimization in water usage and wastage across the textiles supply chain, reduction of waste generation throughout the manufacturing process and the textile supply chain, reduction of carbon emission which impacts in the form of energy across the supply chain in textiles, consideration of biodiversity in terms to preserve and promote with an emphasis on diversification in textile fibers move away from a global dependence on raw materials that utilize unsustainable agricultural practices or result in the depletion of finite natural resources. It also contributes to the textile and fashion processing to minimize pollution, optimize resources, workers and consumers' safety and fulfillments.

Sustainability is one of the key issues to fix above mentioned problems facing by the fashion industry today. The production–consumption relationship needs balance between financial, social and environmental sustainability for indefinite production. Eco fashion or sustainable fashion supports a high degree of environment and social responsibility. In regards, larger trend of sustainable design must appear to be a long-term trend but could last multiple seasons. The practice of sustainability clings on the use of renewable materials and/or non-harmful materials with low-impact processes, reuse or recycling of waste materials, regeneration ability. The next challenge is to make green or environmentally friendly for the sustainable development. It prompts resources rich for all through sustainable consumption and production methods. It creates equity, awareness between public and buyers and the participation of all social groups. Economic and environmental values meet the demand for conservation and resource management.

The possible challenge lies on slow fashion process which ethical practices on design, quality, craftsmanship and experienced labor, the focus on investment and longevity that builds the relationship between retailer and consumer. It develops the following benefits:

- Fair labor
- Enhancing communities
- Satisfying human needs

- Clean and Efficient production
- Cultural diversity
- Sustainable design methods
- Supporting local economies (small scale weavers)
- Diverse business models
- True retail prices
- Robust supply chain relationships
- Meaningful fashion experiences
- Resourcefulness

Aesthetic values in design drive the fashion world and determine how it can eventually be delivered to the consumer. Obstacles in clothing industry start right from fiber production to disposal, transport and CO_2 emission. It gives clear idea about the resources, materials, technologies used in material making, shipment of end products, consumer response on a product, disposal rate and/or impact on end-of-life management of the product. Prior to non-toxic materials minimize impact on the local eco system, emphasize quality and durability over price.

Corporate social responsibility (CSR) and the principles of sustainability influences

a. The consumer behaviour (Social impact of clothing)
b. Sustainable consumption on renewable materials (Supply chain to business and product innovation and consumer engagement)
c. Reducing impact from use

In present times, ethical companies focus on innovation to trend a revolution potentially in the way the fashion industry is run. Exciting things, business models, supply chain systems are added in textiles to drive the sustainable fashion agenda forward and imperative. Both of sustainable and competitive solutions are delivered by pledging innovative trends in slow textiles, future fabrics, re-inventing the supply chain, herbal dyeing and digital printing, creative use of non-fashion materials, vertically integrating production. e.g. LAUNCH 2020

LAUNCH 2020 is a scheme calls for material specialists, experts, designers, producers, entrepreneurs, and organizations from all over the world to collaborate and come up with innovative solutions to the sustainability of the new fabrics. It also introduces upcycling—The performance of finding another use for an existing garment converting (waste) into reusable materials by break down or grinding of high-grade materials into their purest raw forms or substrates. Such a value added activity is necessary to create a product of higher quality or value than the original by which either pre-consumer or post-consumer waste or a combination of the two.

Research focuses on the development of cleaner production technologies to improve the usage of renewable, biodegradable, recyclable materials in toxic free supply chain management.

The Clean Tee—The most sustainable T-shirt on the planet. This has been invented by the company of Nomadix, totally made from recycled textile waste, without adding water or dye. Right First Time (RTF) dyeing the fabrics is in trend by researchers that will improve and limit the wastage of water, energy, dyes and chemicals. Another interesting prone in research is recycling food waste into fiber. In such way citrus is turned into raw materials and afterwards, be spun into yarns.

Algae-Based fabrics attract many customers as well as manufacturers. Algae are fast glowing plants that do not require irrigation as they grow besides rivers, lakes and oceans. Reuse wasted fabrics on textile ground is a prior choice for researchers where in the cutting section to produce another garment. Organic Bamboo fiber is 100% biodegradable fiber that glows fast and without the aid of chemical agent. Eco Spun Recycle employs plastic bottles which have been converted into polyester fiber. Organic cotton fiber is grown without artificial herbicides or pesticides.

Fabric from fermented wine: A group of scientist at the University of Western Australia has produced fabric by culturing bacteria (Acetobacter), letting them to work on wine. Similar work has been done by UK-based BioCoutue which develops a method for creating a lab-grown biomaterial. That is shaped, characterized and even grown into clothing. A fashion designer has developed the material from cellulose bacteria, and dyed in vats of what amounts to sugary green tea. This is lately known as growing garments from microbes [39, 40].

9 Sources of Further Information and Advice

The textile industry is flourishing at high speed for the industrial revolution. With the growing demands of material and sophisticated machines, expansion with latest technologies and that of business, the Supply Chain Management has taken an important role throughout the world. It has taken its own shape in the textile industries for the dynamic growth where large quantities are in demand with varieties of product range. Some of the key components like purchasing decision, choice of disposable textile products and demand for natural fibers are hugely influenced by versatility, flexibility and price offers. High market fluctuations which make the Supply Chain more critical and hence an appropriate management is required in this industry for detoxified products. The cost of natural resources including organic crops, energy in the form of air, water, fuel and network logistics have also direct impact on Supply Chain Management. To overcome this, apparel industry needs to be more systematic. At every stage of manufacturing process, estimation of proper production scale is necessary. At production stage, the technical and commercial aspects are to be taken care of in order to keep the supply chain intact recommended by field experts.

References

1. Rajkishore N, Amanpreet S, Rajiv P et al (2015) RFID in textile and clothing manufacturing: technology and challenges. Fashion and Text 2(9):1–16
2. Smart U, Bunduchi R, Gerst M et al (2010) The costs of adoption of RFID technologies in supply networks. Int J Oper Prod Manage 30:423–447
3. Dharmaraju P (2006) Marketing in handloom co-operatives. Econ Polit Wkly 3385–3387
4. Companies without borders: collaborating to compete. Economist Intell Unit. 2006
5. Hildegunn KN (2004) The global textile and clothing industry post the agreement on textiles and clothing. Discussion paper 5. World Trade Organization
6. The smarter supply chain of the future. Global chief supply chain officer study 1–63
7. Avizit B, Israfil MM, Rifaul Md et al (2014) Supply chain management in garments industry. Glob J Manage Bus Res: A Adm Manage 14(11)
8. Lo CKY, Yeung ACL, Tce Cheng (2012) The impact of environmental management systems on financial performance in fashion and textiles industries. Int J Prod Econ 20(135):561–567
9. http://www.fibre2fashion.com/industryarticle/22/2142/airdye-technologycoloringtextiles-without-the-use-ofwater1.asp
10. Vijaya C, Poovendhiran NV (2009) A study on the factors determining the business success and failure of small scale industry units with reference to erode city. J Contemp Res Manage 21–29
11. Alper S (2008) The U.S. fashion industry: a supply chain review. Int J Prod Econ 114:571–593
12. Shui S, Wohlgenant MK, Beghin JC et al (1993) Policy implications of textile trade management and the U.S. cotton industry. Textile trade policy implications 37–47
13. http://www.cottonconnect.org/what-we-do/introduction.aspx
14. Sami AM (2013) Technology of eco friendly textile processing—A route to sustainability 5 Mar 2013
15. Sarwar MI, Ali MA (2013) Sustainable and environmental friendly fibers in textile fashion-A study of organic cotton and bamboo fibers. Report No. 2010.9.14, 5 Mar 2013
16. https://www.icat.vt.edu/sites/default/files/3D%20Printing%20Textile%20Solutions.pdf
17. Ascloy N, Dent K, Haan ED (2004) Critical issues for the garment industry. Somo bulletins on issues in garments & textiles 1–88
18. https://www.scribd.com/doc/38255347/Promotion-Strategy-of-Vishal-mega-Mart
19. www.cleangredients.org
20. Luongo G. (2015) Chemicals in textiles—A potential source for human exposure and environmental pollution. Doctoral thesis, 1–53
21. Hasanuzzaman Bhar C (2016) Indian textile industry and its impact on the environment and health: a review. Int J Inf Syst Serv Sect Arch 8(4):33–46
22. http://www.medicaldaily.com/6-carcinogenic-flame-retardants-found-bodies-and-homes-california-residents-310624
23. Baek YJ, Shin J (2014) Risk points of flame retardant textiles by halogen and halogen-free laminating film. Mater Sci Appl 5:830–836
24. Travaglini R (2000) Human health risk assessment J "Tribufos". U.S. Environmental protection agency office of pesticide programs health effects division
25. Bharadwaj A, Yadav D, Varshney S (2015) Non-biodegradable waste—its impact & safe disposal. Int conference on Technologies for sustainability—Engineering, information technology, management and the environment 391–398
26. Gonias ED, Oosterhuis DM (2004) Effect of the plant growth regulator PGR-IV Plus as a safener for glyphosate applications in cotton. Summaries of Arkansas Cotton Res 72–75
27. RCRA in focus: textile Manufacturing 1–15
28. Brito MPD, Carbone V, Blanquart CM (2008) Towards a sustainable fashion retail supply chain in Europe: organization and performance. Int J Prod Econ 114:534–553
29. Guidelines for GOTS certification for textiles. OneCert Asia 1–9

30. Prusty A, Gogoi N, Jassal M et al (2010) Synthesis and characterization of non-fluorinated copolymer emulsions for hydrophobic finishing of cotton textiles. Indian J of fibre and textile research 35:264–271
31. http://www.global-standard.org/the-standard/general-description.html
32. https://www.nibusinessinfo.co.uk/content/import-regulations-textiles-sector
33. Roznev A, Puzakova E, Akpedeye F et al. Recycling in textiles. Supply chain Management. 1–20
34. Ernst C, Ferrer AH, Zult D (2005) The end of the Multi-fibre arrangement and its implication for trade and employment. Employment strategy papers
35. http://organicclothing.blogs.com/my_weblog/health_wholeness/
36. Rajendran Nair K (1996) India's handloom sector @ http://www.pib.nic.in. Best management practices for pollution prevention in the textile industry. Environmental protection agency, US
37. World Trade Organization (WTO) (2001) Accession of the People's Republic of China—Decision of 10 November, WT/L/432, Annex 5B, 23 November 2001, Geneva
38. Pioneering a sustainable textile industry. Sustainability innovation Collaboration. Ppt
39. Eryuruk SH (2012) Greening of the textile and clothing industry. Fibres Text Eastern Europe 6A(95):22–27
40. http://www.ssfindex.com/results-2012/

Environmental Issues in Textiles: Global Regulations, Restrictions and Research

Harinder Pal

Abstract Textile industry is facing many criticisms due to huge amount of water consumption and hazardous nature of various chemicals being used throughout from raw material extraction till disposal phase. The industry has been responsible for creating land, air and water pollution affecting human health and damaging the eco-system. Chemicals, used during cultivation and different wet processing phases get hold to the textile, evaporate into air and drain through water. They are absorbed through skin, inhaled through air and swallowed through food and water. Chemicals, being toxic and carcinogenic trigger serious diseases and allergic reactions in people. Energy consumption and heat emission are other factors disturbing the eco-system. Ecological considerations are becoming important factors all over the world. To protect humans and the environment from harmful substances, various countries have introduced stringent legislations. It is becoming mandatory to comply with environmental regulations and violation of the provisions can be prosecuted as a criminal offence. This chapter discusses the key environmental issues involved in textile industry, restricted chemicals and substances and their impact on human health, regulation imposed along with research aspects which are being applied to curb the harmful impacts and to craft the textile industry more sustainable.

Keywords Environmental issue · Air pollution · Water consumption
Water pollution · Restricted chemicals · Azo dyes · Heavy metals

---

The original version of this chapter was revised: For detailed information please see Erratum. The erratum to this chapter is available at https://doi.org/10.1007/978-981-10-4777-0_4

H. Pal (✉)
Department of Fashion Technology, B. P. S. Mahila Vishwavidyalaya,
Sonipat, Haryana, India
e-mail: aroraharinder@yahoo.com

27

1 Introduction

Textile industry has a prominent role in global economy but at the same time it has been condemned as one of the most polluting industry having inconsiderate effect on environment and human health. Today, the industry is facing many criticisms due to huge amount of water consumption and hazardous nature of various chemicals being used throughout from raw material extraction till disposal phase. Chemicals, used during cultivation and different wet processing phases get hold to the textile, evaporate into air and drain through water. They are absorbed through skin, inhaled through air and swallowed through food and water. Chemicals, being toxic and carcinogenic trigger serious diseases and allergic reactions in people. Energy consumption and heat emission are other factors disturbing the eco system. Garment industry emits a lot of heat and carbon which is equally responsible for carbon emissions into the environment and for the ozone depletion [1]. So the whole textile industry has been responsible for creating land, air and water borne hazards affecting human health and damaging the eco system. There is a dire need to act more with environmental responsibility from producer, consumers, and policy makers. Lack of information and knowledge in an attempt to produce products in an unsustainable way creates uncertainty about the safety aspects from health point of view and as well as to the environment. Though the environmental legislations have been made stringent in many countries yet there is a need to enforce it strictly. Also rather than being policy driven, environmental concern is a social responsibility for which the contribution is required not only from the industry but also from the consumers. Consumer needs to be aware about his role to protect the environment by purchasing eco-friendly product. Similarly the manufacturer must be aware of the harmful impact of the chemicals and processes being used to safeguard the interest of the human health and the environmental concerns. This chapter discusses the key environmental issues involved in textile industry, restricted chemicals and substances and their impact on human health, regulation imposed along with research aspects which are being applied to curb the harmful impacts and to craft the textile industry more sustainable.

2 Textile Industry and Its Impact on Environment

Textile industry covers a wide spectrum of manufacturing activities and is diverse in terms of raw materials used, techniques employed, chemicals used and the final products [2]. It has been responsible for creating land, water and air pollution all along its supply chain from raw material extraction till disposable phase.

Raw Material cultivation requires land for crops; consumes energy and water; removes nutrients from soil; and pollutes water by using chemicals like pesticides, biocides and herbicides. Chemicals like defoliants spread in air and are dangerous to human health. Synthetic fibre production also consumes valuable resources like petroleum, coal and oil. Spinning, weaving and garmenting release solid waste in the form of fibre, yarn, fabric off-cuts, and dust. The loose fibres and micro dust if inhaled can affect our respiratory system. These activities also employ chemicals like delustering, anti-static and other finishes. These manufacturing activities create noises, discharge smoke and effluents. These operations consume resources like energy and water which are limited reserve. Wet processing sector consumes huge amount of water and is responsible for creating waste water effluents. Water is contaminated by detergents, soaps and bleaches. Finishing process gives birth to toxic by-products and gases from chemicals, dyes and resins. It has major impact on human health and unbalancing the eco system, the detail of which is covered subsequently in the chapter. The major operations performed in wet processing include various operations like desizing, scouring, mercerizing, bleaching, dyeing, printing and finishing. Cleaning and maintenance also make uses of harmful and strong chemicals. Process of distribution pollutes the air with fuel, which is not an infinite supply for distribution [3].

In general, textile industry generates three kinds of waste i.e. solid wastes, liquid effluents and air emissions. Quantity and nature of waste generated depends on the fabric being processed, chemicals being used, technology being employed, operating practices etc. Apart from it, various utilities in the textile units like water treatments plants, workshops, cooling towers, boilers, thermo pack, diesel generators etc., are also responsible in generating wastes [4]. To what extent, these various segments have been responsible in damaging the eco system is mentioned subsequently.

2.1 Impact of Cotton Production [5, 6]

Cotton is the main ingredient of denim. Cotton growing requires the use of pesticides, insecticides, herbicides which in turn impose hazards on human health and the environment. Some of the environmental impacts associated with improper use of these chemicals are:

- Contamination of water in rivers, ponds, groundwater etc.
- Impact on pattern of crop rotation
- Poisoning of aquatic living species
- Loss of bio-diversity
- Poisoning of livestock used to feed cattle and detected in milk and meat causing serious diseases
- Loss of soil's nutritional value
- Inhalation of contaminated air by human and other living beings causing diseases.

2.2 Water Consumption

Water, being the most prestigious and scarce resource of the earth is used in enormous scale in the textile industry. Natural fibre cultivation especially cotton and textile processing consumes a huge amount of water making the whole industry responsible for high water footprint. According to research done by Water Footprint Network, producing 1 kg of cotton in India, it consumes 22,500 litres of water on average. The global average water footprint for 1 kg of cotton is 10,000 litres. The far higher water footprint for India's cotton is due to inefficient water use and high rates of water pollution because of huge amount of pesticides used in cotton production. By exporting more than 7.5 million bales of cotton in 2013, India also exported about 38 billion cubic metres of virtual water. It was estimated to be enough to supply 85% of the country's 1.24 billion people with 100 litres of water every day for a year [7, 8]. Even the denim industry, which is an important subsector of the textile industry, is on top of the pyramid of water consuming textiles with a total volume of 2900 gallons (approx. 11,000 litres) consumed per pair of jeans 9–12]. As per Luiken et al., the production of jeans is estimated to be more than 3.5 billion pairs. If an average pair of jeans weighs 600 g, the total textile consumption of jeans is above 2.1 million metric tons a year. Considering water consumption 11,000 litres per pair of jeans and production to be 3.5 billion pairs, water resource consumption to the extent of 38.5 trillion litres or 38.5 billion cubic meters will be required [13].

The mechanical operations like spinning, weaving consumes very less water as compared to textile wet processing operations, where water is used extensively. In textile wet processing, water is used mainly as a solvent for processing chemicals as well as for washing and rinsing medium. Apart from this, some water is consumed in ion exchange, boiler, cooling water, steam drying and cleaning. The quantity of water required for textile processing varies from mill to mill depending on fabric produce, process, equipment type and dyestuff. The consumption of water for dyeing process in a typical cotton mill can be as high as 300,000 litres/1000 kg of product and for the product made from synthetic fibre, dyeing may consume 17,000–34,000 litres/1000 kg. Scouring process may even consume 20,000–45,000 litres/1000 kg of cotton products [14].

2.3 Water Pollution

Use of fertilizers, pesticides with large quantities of water for cotton production has affected large scale ecosystems, which has impacted the health and wellbeing of people living there. The chemicals used during textile processes, which include mainly dyes and pigments, detergents, surfactants, are the main sources of water pollution. The waste water released contains high levels of salts, acids or alkali chemicals, surfactants, pH concentration and dye colours. The effluents released may contain toxic substances which affect aquatic life as well as other species including human being. Several non-process chemicals used in machine cleaners, boiler treatments, clean ups etc. can also add up the pollution load.

The presence of colour, dissolved solids (TDS), suspended solids, toxic (heavy) metals, and residual chlorine in the effluent results in high chemical oxygen demand (COD). The presence of organic pollutants leads to high bio-chemical oxygen demand (BOD). If necessary steps are not taken by the denim industry, polluted wastewater can cause serious environmental problems [5, 15].

2.4 Textile Wastes [2]

Textile industry is a major source of wastes generated through many production processes like spinning, weaving, knitting, finishing, and garmenting. Wastes generated in textile industry can be classified into three categories:

2.4.1 Hazardous or Toxic Wastes

These include colours, metals, phenol, toxic organic compounds, phosphates, chlorinated solvents etc. Some of these wastes can also come from non-textile processes such as machine cleaning, boiler chemicals etc. Certain non-biodegradable organic materials like surfactants and solvents in the wastes resist biological effluent treatment process and produce aquatic toxicity when discharged into downstream.

2.4.2 Dispersible Wastes

The sources of dispersible wastes are wide spread in textile wet processing. These include print pastes, wastes from back coating operations, batch dumps of unused process chemicals etc.

2.4.3 High Volume Wastes

These include wash water from preparatory, dyeing and printing operations and the exhausted dye baths.

2.5 Air Pollution in Textile Industry [2]

Textile industry apart from creating land and water pollution is also responsible for creating air emissions through spinning, napping, shearing, drying and finishing operations. Air emissions include dust, oil mists, acid vapours, odours and boiler exhausts which are accountable in creating air pollution. Air pollution is the introduction of chemicals, particulate, or biological materials that cause harm or discomfort to humans or other living organisms, or damage the natural environment into the atmosphere. Gaseous emissions have been identified as the second greatest pollution problem (after effluent quality) for the textile industry. The major air pollution problem in the textile industry occurs during the finishing stages, where various processes are employed for coating or treating the fabrics with various finishes. Coating materials include lubricating oils, plasticizers, paints and water repellent chemicals–essentially, organic compounds such as oils, waxes or solvents. After the coating operations, the coated fabrics are cured by heating in stenter, ovens or dryers which results into vaporization of high molecular weight volatile organic (usually hydrocarbon) compounds (VOCs). Other air emissions may include formaldehyde, acids, softeners and other volatile compounds. Carriers and solvents may be emitted during dyeing operations depending on the types of dyeing processes used and from wastewater treatment plant operations. Nitrogen and sulphur oxides are also generated from boilers. Carriers used in batch dyeing of disperse dyes may lead to volatilisation of aqueous chemical emulsions during heat setting, drying, or curing stages. Acetic acid and formaldehyde are two major emissions of concern in textiles [4, 16]. Various kinds of air pollution in the textile industry can be categorized in four segments [2]:

– Oil and acid mists
– Solvent vapours
– Odour
– Dust and lint

2.5.1 Oil and Acid Mists

Spinning oils, plasticizers etc. used during the processes get evaporated or degrade thermally when subjected to heat. Acids mists are also produced due to volatilization of organic acid like acetic acids etc. These mists are corrosive in nature.

2.5.2 Solvent Vapours

Solvent vapours consist of large number of toxic chemicals in varying concentration depending upon the chemical compounds used during dyeing, printing and finishing operations. Some of these compounds include formaldehyde, chlorofluorohydrocarbons, mono and dichloro-benzene, ethyl acetate, hexane, styrene etc.

2.5.3 Odour

Odour is another aspect of air pollution which is associated with oil mists or solvent vapours. It affects the aroma senses in the human beings. Various sources of such problems are resin finishing, sulphur dyeing of cotton, polyester dyeing, dye reduction or dye stripping with hydrosulphite and bleaching with sodium hypochlorite etc.

2.5.4 Dust and Lint

Fly generation in the form of dust and lint are produced during processing of natural and synthetic fibres prior to and during spinning, napping, carpet shearing etc. Inhalation of these micro particles is associated with many respiratory diseases.

3 Regulations, Chemical Restrictions and Associated Health Hazards

Ecological considerations are becoming important factors all over the world. To protect humans and the environment from harmful substances, various countries have introduced stringent legislations. It is becoming mandatory to comply with environmental regulations and violation of the provisions can be prosecuted as a criminal offence. The impact of the regulations has been strong on textile exporting countries and indicates its future implications too. European Union has taken a lead by imposing a stringent legislation to protect the environment giving emphasis on sustainable practices. It takes into account the whole environmental performance of the plant, including emissions to air, water and land, generation of waste, use of raw materials, energy efficiency, noise, prevention of accidents and risk management. The objective of the legislation is to protect human health; preserve, protect and improve the quality of the environment; sensible use of natural resources etc. European Union (EU) environmental legislation comprises large number of detailed legal acts of relevance for EU environmental policy [17].

A large number of chemicals in vast quantity used throughout the various processes have resulted into harmful impact on human beings. Some of the pesticides, insecticides and fertilizers used during cotton cultivation have resulted into developing serious diseases among living beings. Huge amount of pollutants are released during wet processing stages especially sizing, scouring, bleaching, dyeing, finishing/washing and rinsing processes. Dye bath effluents may contain heavy metals, ammonia, alkali salts, toxic solids and large amounts of pigments—many of which are toxic. Colorants may contain organically bound chlorine, a known carcinogen [1]. Many of the health problems which have been reported or asserted by the use of hazardous chemicals and other substances include [18]:

- Excessive use of formaldehyde may cause skin irritations.
- Presence of toxic residue e.g. 'lindane' in the blood due to contamination of textiles by the use of pesticides in cotton farming like Lindane, DDT and hexachlorocyclohexane.
- Poisonous—toxic residue of preserving agents on cotton and wool.
- Risk from carcinogenic dyes/azo dyes.
- Sensitizing (allergic effects) from azo dyes, formaldehyde, optical brighteners, and softeners.

Various governments and private organisations have restricted a number of substances looking at their potential negative effects on the environment and human health and are therefore regulated. European legislation has introduced a single integrated registration system (REACH) and a Globally Harmonized System of Classification and Labelling of Chemicals (GHS) to ensure that dangerous substances are identifiable [17].

List of chemical compounds which are found to be hazardous for human health and are restricted in many countries includes:

– Azo/Carcinogenic/Allergenic Dyes
– Formaldehyde
– Toxic pesticides
– Pentachlorophenol
– Heavy metal traces
– Halogen carriers
– Chlorine bleaching etc.

3.1 Azo/Carcinogenic/Allergenic Dyes

Azo dyes have the potential to split into the arylamines from which they were synthesized by chemical reduction (e.g. dye stripping) or enzymatic processes in the metabolism of organisms. This release of carcinogenic amine through reductive splitting from Azo dyes is the major reason to keep it in potentially carcinogenic category. The most prominent examples are dyes based on benzidine and congeners [18].

24 aromatic amines have been confirmed as, or implicated to be, carcinogens in humans. They can have acute to chronic effects upon organisms, depending on exposure time and Azo dye concentration. Some aromatic amines, particularly benzidine, 2-naphthylamine, and 4 aminobiphenyl, dramatically elevates bladder cancer risk. In one German dye plant, 100% of workers (15) involved in distilling 2-naphthylamine developed bladder cancer [21].

1,4-diamino benzene is an aromatic amine whose parent azo dyes can cause skin irritation, contact dermatitis, chemosis, lacrimation, exopthamlmose, permanent blindness, rhabdomyolysis, acute tubular necrosis supervene, vomiting gastritis, hypertension, vertigo and, upon ingestion, oedema of the face, neck, pharynx, tongue and larynx along with respiratory distress. Because of their chemical stability and synthetic nature, reactive Azo dyes are not totally degraded and exhibit slow degradation by conventional wastewater treatment methods. Aromatic amines can be mobilised by water or sweat, which aids their absorption through the skin and other exposed areas, such as the mouth. Absorption by ingestion is faster and so potentially more dangerous, as more dye can be absorbed in a smaller time frame. Water soluble Azo dyes become dangerous when metabolised by liver enzymes [21].

Till 1994, there were no legal regulations in nearly all countries for the prohibition of their production, import or use of carcinogenic dyes and those which degrade to carcinogenic arylamines. In Germany, in July 1994, there were amendment to the ordinance of consumer goods to legally prohibit the use of azo dyes capable of releasing carcinogenic arylamines through reductive splitting of azo bonds, as well as prohibiting the import and trade of textiles dyed with them. However the deadline for manufacturing and importing textiles containing these dyes into Germany was March 31, 1996 and for trade was September 1996 [18]. Following that few other countries also imposed similar restrictions. These include Netherlands, Turkey, France and India. However, the approach in enforcing ban varied from one to another [19].

The European Commission has laid out its stance on Azo dyes in Section 43 (Azo dyes and Azo colourants) of Annex XVII of REACH. Following Aromatic Amines have been restricted from European Union [21].

- 4-aminodiphenyl/xenylamine/Biphenyl-4-ylamine (CAS no. 92-67-1)
- Benzidine (CAS no. 92-87-5)
- 4-chloro-o-toluidine (CAS no. 95-69-2)
- 2-naphthylamine (CAS no. 91-59-8)
- o-aminoazotoluene/4-o-tolylazo-o-toluidine/4-amino-2',3dimethylazobenzene (CAS no. 97-56-3)
- 2-amino-4-nitrotoluol/5-nitro-o-toluidine (CAS no. 99-55-8)
- p-chloranilin/4-chloroaniline (CAS no. 106-47-8)
- 2,4-diaminoanisole/4-methoxy-m-phenylenediamine (CAS no. 615-05-4)
- 4,4'-diaminodiphenylmethane/4,4-methylenedianiline (CAS no. 101-77-9)
- 3,3'-dichlorobenzidine/3,3'dichlorobiphenyl-4,4'-ylenediamine   (CAS   no. 9194-1)
- 3,3'-dimethoxybenzidine/o-dianisidine (CAS no. 119-90-4)
- 3,3'-dimethybenzidine/4,4'-bi-o-Toluidine (CAS no. 119-93-7)
- 3,3'-dimethyl-4,4'-diaminodiphenylmethane/4,4'-methylenedi-o-toluidine (CAS no. 838-88-0)
- p-cresidin/6-methoxy-m-toluidine (CAS no. 120-71-8)
- 4,4'-methylene-bis-(2-chloro-aniline)/2,2'-dichloro-4,4'methylenedianiline (CAS no. 101-14-4)
- 2,4-Xylidine (CAS no. 95-68-1)
- 2,6-Xylidine (CAS no. 87-62-7)
- 4,4'-oxydianiline (CAS no. 101-80-4)
- 4,4'-thiodianiline (CAS no. 139-65-1)
- o-toluidine/2-aminotoluene (CAS no. 95-53-4).

Sensitizing (allergenic) dyes have also been included in Oeko-Tex Standard-100 as a required parameters which in most of the cases belongs to disperse dyes. Some of the disperse dyes such as C.I. Disperse Yellow 3, Disperse Orange 3, Disperse Red 1 and 17, Disperse Blue 3, 106 and 124 as well as Naphtol AS comes under this category. These dyes can migrate out of the fibers and cause harm to the skin [18].

3.2 Formaldehyde

Formaldehyde in textile industry is used to produce crease resistant properties and to prevent shrinkage. It acts as a cross-linking agent to make an easy-care finish. Formaldehyde is a toxic chemical and the release of it can be harmful to health through irritation of mucous membranes and the respiratory tract. It is restricted by law in some countries or by voluntary specifications of textile producers. As per Oeko-Tex standard-100, Formaldehyde concentrations are tested in accordance to the test specification of the Japan Law 112. The minimum detectable level in test procedure is generally 20 ppm which is also the minimum specified limit kept for baby wear. The permissible limit till 75 ppm is applicable for clothing worn directly on skin like underwear, shirts, blouses and stocking etc. and the highest permissible limit 300 ppm is for outwear apparel [18, 20].

3.3 Pesticides

Pesticides are used for cotton cultivation to prevent it from insects and also used as a moth protection during storage. Herbicides are also used for weed-eradication and as defoliant chemicals. Pesticides and herbicide residues are rated slightly to strongly toxic causing potential health hazards. The toxic substances can be absorbed by the fibres and might remain in the final product and sometimes easily assimilated through the skin. In one of the case study, 'lindane' was found in the human blood, may be due to contamination of textiles by the use of pesticides. The pesticides mentioned in the Oeko-Tex Standard-100 table are chlorinated substances and their use is restricted in most of the cotton growing countries. Lower limit of 0.5 ppm is for baby articles [18, 20].

3.4 Pentachlorophenol

Chlorinated phenols like Pentachlorophenol (PCP) are used as preserving agents for textiles to prevent mold spots caused by fungi. These are toxic in nature and regarded as cancer-inducing agents. According to German regulations, the production and the use of PCP is banned, and impurities in textiles and leather products are allowed up to 5 ppm. However Oeko-Tex standard-100 prescribes the maximum limit at 0.5 ppm [18, 20].

3.5 Heavy Metals

Heavy metals are constituents of some dyes and pigments, which may be introduced into textiles through dyeing and finishing processes. They can also exist in natural fibres due to absorption by plants through soil [20].

The use of heavy metals is common in textile processing and finds various applications like in dyes, fastness improvers, oxidizing agents for sulphur and vat dyes etc. Chromium, cobalt, copper and nickel compounds are used in metal complex dyes. Copper compounds are used in fastness improvers. Potassium dichromate is used in after chroming of mordant dyes on wool [18].

Some of the heavy metals, which may exist in textile or leather industry are: Antimony (Sb), Arsenic (As), Lead (Pb), Cadmium (Cd), Mercury (Hg), Copper (Cu), Chromium (Cr), Cobalt (Co), Nickel (Ni).

These heavy metals if absorbed by humans tend to accumulate in internal organs such as the liver or kidney and can be serious threat to the human health. For example, high level of lead can seriously affect the nervous system. Cadmium and lead are classified as carcinogens. Both are restricted in US under Consumer Product Safety Improvement Act (CPSIA) and in Europe as per European regulations [20].

Chromium mainly releases during leather tanning process when chrome tanning is employed. It is strong oxidant and is carcinogenic. Nickel is found in alloys used for metal accessories on garments such as buttons, zippers and rivets. Some people are allergic to nickel and may experience skin irritation when in contact with nickel for an extended period. The release of Nickel is restricted under the EU REACH Regulation (EC) No. 1907/2006, Annex XVII [20].

As per Oeko-Tex Standard 100, the use of metal complex dyes is not prohibited. Restriction imposed on harmful metals refers to the maximum limit of concentration of extractable harmful metals on textiles rather than total amount of metals on the textiles when dyed with metal complex dyes. Some of the metal complex dyes which use copper or nickel phthalocyanates result into producing important shades like turquoise or brilliant green. Premetallized or afterchroming dyes also improve wet fastness in dyeing of wool or polyamide. Metal complex dyes also result into improved light fastness in case of violet, blue or navy reactive and direct dyes. However in case of direct dyes having metals poses problems because of their limited wet fastness, so these dyes cannot achieve the desired specified limits in all the cases. In case of wool dyed with mordant dyes, traces of chromium are leached by perspiration solutions but to a greater extent by saliva solutions [18].

3.6 Alkylphenols (AP) & Alkylphenol Ethoxylates (APEO)

Alkylphenols and alkylphenol ethoxylates are commonly used as wetting agents in textile processing. EU REACH Regulation (EC) No 1907/2006 restricts the discharge of Nonylphenol (NP) and Nonylphenol Ethoxylates (NPEO). NPEO's have been used as detergents, emulsifiers, wetting agents and dispersing agents for many years. NP is the intermediate to synthesize NPEO. NPEO and NP are very toxic to aquatic life and considered aquatic pollutants. They can disrupt the hormone-regulating system of aquatic animals and cause estrogenic effects. Octylphenol (OP) and Octylphenol Ethoxylates (OPEO) are the other AP and APEO's commonly concerned [20].

3.7 Perfluorooctane Sulfonates (PFOS)

PFOS are widely used to provide grease, oil and water resistance to textiles, apparel, carpets, leather and paper. The substance is considered to be very bio-accumulative and toxic [20].

3.8 TBT, DBT and Other Organotin Compounds

TBT has been used for preventing the bacterial degradation of sweat to avoid unpleasant odour of socks, shoes and sports clothes. Some organotin compounds may be used in PU and PVC productions. These compounds when used in higher concentrations are harmful and toxic in nature. These are absorbed through skin and are suspected to cause reproductive disorders [20].

3.9 Halogenated Carriers

Carriers are used as auxiliaries for dyeing of polyester fibers with disperse dyes at atmospheric pressure and temperatures below 100 °C. The highly effective and inexpensive halogenated types such as trichlorobenzene and dichlorotoluene are accused of being toxic and harmful to the environment [18]. These can have adverse effects on central nervous system. These may induce liver and kidney malfunction [20].

Alternatives are carriers based on aromatic carbonic acids and alkyl phthalimides with better toxicologic properties but higher prices or, if possible, dyeings without carriers under high pressure or through continuous processes [18].

3.10 Flame Retardants

TRIS, TEPA, 2,3-Dibromopropyl Phosphate, Polybrominated Biphenyls (PBB) and Polybrominated Diphenylether (PBDE) are commonly used flame retardants.

Prolonged contact to high dosages of flame retardants can cause impairment of the immune system, hypothyroidism, memory loss and joint stiffness [20].

4 Eco Label And Certifications [20]

To meet the international regulations and environmental standards, Eco Labels and certification schemes are proposed by government and private enterprises to minimize the adverse environmental impacts from the design through installation, usage and disposal stages of a product. In order to sustain and compete in the global market, companies are maintaining norms to be eligible for Eco Label and getting certification to show their concern towards environment and to woo the customers. Labels also allow consumers to make comparisons among products and give consumers the ability to reduce the environmental impacts of their daily activities by purchasing environmentally preferable and healthy products and minimising their consequences during use and disposal.

5 Pollution Control Techniques [2]

5.1 Cleaner Production Techniques and Processes

This involves various integrated approaches to reduce water and raw material consumption, substituting non-toxic chemicals, recovery of chemicals, waste minimization and elimination, energy conservation, optimum use of resources along the entire product life cycle. Techniques involved must be evaluated and modified continuously in terms of choice of process, their sequence and the equipment used.

5.2 End-of-Pipe Treatments

This refers to use efficient effluent treatment strategy. Rather than treating combined and complex effluents, segregating treatment of specific effluent streams must be adopted. Different kinds of residue in the form of sludge (from biological or physic-chemicals units) and waste materials (from production process) requires proper disposable alternatives such as compaction, land filling and anaerobic digestion or incineration.

5.3   Sludge Treatment and Disposal

Sludge formation if released in small quantity during biological treatment of effluent can be recycled otherwise aerobically digested methods are used and subsequently thickened by gravity thickener. Supernatant is returned to aeration tank and thickened sludge may be de-watered. Sludge from the centrifuge may be disposed off to a sanitary landfill facility on or off site.

5.4   Solid Waste Disposal

Solid waste from textile industry may include cans, rejected fabric pieces, dust, fibres or yarn waste etc. These wastes may be sent for either recycling or incineration or may be carted away to landfill sites.

5.5   Air Emission Control

The measures to control air pollution depend upon the kinds of emission which may be in the form of particulate, scrub fuel gases consisting oil mists, volatile organic carbon (VOC), organic solvent vapours etc. Particulate emissions are generally controlled by cyclone separators, bag filters and wet scrubbers.

Oil mists and volatile organic carbon (VOC) are more difficult to control. Proper air ducting and mist eliminators are used to control pollution. Oil mist eliminators involve various techniques and steps like:

– Pre-removal of dust and lint by fabric filters.
– Condensation of vapours by cooling the contaminated air by direct contact cooling or heat recovery via heat exchange. Techniques generally used for direct contact cooling are low energy scrubbers, spray towers and packed towers.
– Mist removal from air either by electrostatic precipitator or using high efficiency fibre mist eliminators when oil mists contain excessive water to avoid arcs and short circuit. In this case, no pre-cooling or condensation is necessary.
– Collection and disposal of contaminant is also achieved by ducting to chimney with adequate dispersal height.

Organic solvent vapours, which are released during drying, finishing and solvent processing operations, have limited solubility in water. So these vapours cannot be treated by scrubbing. Incineration is expensive technique and the emissions exhausted also need to be handled separately. Use of activated carbon for vapour adsorption is effective strategy apart from solvent recovery techniques.

There are regulations to use specific type and composition of fuel and the minimum chimney height for satisfactory pollutant dispersal. Efforts must be done to reduce the oil mists and VOC by controlling the application of spinning oils and finishing agents to fabric.

6 Identification of Chemical Residues in Textiles [19]

To ascertain the presence of restricted chemicals and substances within the prescribed limit, it becomes necessary to adopt high end instrumental analysis for their presence and identification in the product. Chromatographic and spectroscopic equipments are used.

6.1 Testing of Banned Amines

The detection of banned azo dyes used in synthetic fibres requires extraction of sample using an appropriate organic solvent. The solvent used may be formic acid in case of nylon, dimethyl formamide in case of polyacrylate and xylene in the case of polyester. The extract is then analysed by any of the following chromatographic techniques:

- Gas Chromatography with Mass Spectrometer (GC-MS)
- High Performance liquid chromatography with Diode Array Detector (HPLC-DAD)
- High Performance Thin Layer Chromatography (HPTLC)
- High Performance Capillary Electrophoresis with Diode Array Detector (HPCE-DAD).

It is suggested that carcinogenic amines must be estimated by using at least two different methods in order to avoid any possible misinterpretation caused by interfering substances. However the quantification of the banned amines may be done using HPLC with DAD.

6.2 Testing of Heavy Metals

Heavy metals present in the textile products are extracted by artificial saliva or perspiration solutions in a liquor to material ratio of 20:1 at 40 °C for one hour [2]. Alternatively, the textile sample is digested using concentrated nitric acid over a hot plate or microwave digestion system. This extract or digested solution is analysed by using either atomic absorption spectrometer (with flame and graphite furnace) or inductively coupled plasma emission spectrometer. The level of accuracy and minimum detection limit are better with atomic absorption spectrometer (AAS) but the technique is not preferable for the simultaneous analysis of 4 or more elements.

In such cases, inductively coupled plasma emission spectrometer is used. In AAS, cathode of the lamp is coated with the metal of interest for the analysis of that element and maximum three lamps are required for the analysis of maximum of three elements.

While operating on Atomic absorption spectrometer (AAS), the extract is introduced into the flame or graphite furnace. Chemical bonds of the sample are broken and completely atomized when exposed to high temperature. Radiation emitted by hollow cathode lamp, the cathode of which is coated with the metal to be analysed, are made to pass through the vapour layer of atoms. The detector measures the decrease in the light intensity of radiation emitted by the lamp, which is a measure of the quantity of the tested metal.

6.3   Testing of Formaldehyde

Testing of formaldehyde are generally carried out by any of following three methods:

6.3.1   Japanese Test Method 112

In this method, extraction from the specimen at 40 °C is made to react with acetyl acetone and its quantification is measured at 412 nm using uv-visual spectrometer.

6.3.2   BS:6806

In this method, chromotropic acid is used for colour development and the absorption is measured at 585 nm using uv-visual spectrometer.

6.3.3   AATCC Method112

In this method, formaldehyde evolved from the textile under the accelerated storage condition is determined. The extraction and measurement method is similar to Japanese Method.

6.4   Testing of PCP

The sample extracted with methanol solution is subjected to a multi-step cleaning operation to eliminate the contaminants and then acetylated with acetic anhydride. The acetylated phenols in the extract are separated by capillary column in the gas chromatograph (GC) and are then detected and quantified by electron capture detector (ECD).

6.5 Testing of Pesticides

The sample extracted with suitable solvent is cleaned up over gel permeation chromatography (GPC) to eliminate interfering matrix compounds of higher molecular weight. It is then analysed by the GC with electron capture detector (ECD) or mass spectrometer (MS).

6.6 Testing of Other Quality Characteristics

Other quality parameters like colour fastness are tested as per standard methods of testing, viz. AATCC, BS, ISO etc. This becomes important as the bleeding of colour due to washing, perspiration, rubbing can be a cause of water effluents discharging chemicals exceeding norms limits.

7 Guidelines and Suggestions

Various eco-friendly considerations which may help to move towards sustainability are:

- Efficient use of resources like water, energy, chemicals, labour
- Use of naturally coloured cotton, organic/green cotton, natural and eco friendly dyes, auxiliaries and packaging materials etc.
- Promoting Eco Standards and Eco Labels
- Environmental monitoring exercise: measuring liquid and gas flows in channels, analysis of solid, liquid and gaseous samples, Analysis of VOC in air from finishing operation, noise levels during weaving on loom, wet analysis such as BOD, COD and suspended solids etc. to assess the impact on environment, health and safety.

8 Conclusion

Textile Industry has impacted a lot to the environment and human health. Ecological issues are becoming important. Regulations are being enforced to curb the negative impact which the textile industry has done to the society.

Manufacturing industries are responding but at a slower pace. It needs to adopt the processing techniques and eco friendly products to comply with environmental regulations. To ensure the eco-friendliness, they should adopt appropriate measures for product evaluation, process evaluation and validation for their eco-friendly features. In other words, they need to adopt a "cradle to grave approach" for the manufacture of eco friendly textiles.

Acknowledgements The Information has been collected from various sources and the author has tried his best to acknowledge them. Author expresses his apologies if someone is left by chance and due regard is given to all. The author is thankful to Intertek Testing Services, IJFTR, AATCC and Parliamentary office of Science and Technology, UK for their valuable contribution.

References

1. Ballikar V (2017) Textile industry and environmental issues (online). Available: http://www.fibre2fashion.com/industry-article/6785/textile-industry-and-environmental-issues. Accessed 11 Jan 2017
2. Chavan RB (2001) Indian textile industry-environmental issues. Indian J Fibre Text Res 26:11–21
3. Fibre2fashion (2017) Textiles in an environmental perspective, Available (online): http://www.fibre2fashion.com/industry-article/49/textiles-in-an-environmental-perspective. Accessed 11 Jan 2017
4. Tiwari M, Babel S (2013) Air pollution in textile industry. Asian J Environ Sci 8(1): 64–66
5. Amutha K, (2017) Environmental impacts of denim. In: Sustainability in denim, Elsevier Publications. http://dx.doi.org/10.1016/B978-0-08-102043-2.00005-8
6. ICAC (2015) Measuring Sustainability in Cotton Farming Systems (online): https://www.icac.org/getattachment/Home-International-Cotton-Advisory-Committee-ICAC/measuring-sustainability-cotton-farming-full-english.pdf
7. https://www.theguardian.com/sustainable-business/2015/mar/20/cost-cotton-water-challenged-india-world-water-day
8. http://waterfootprint.org/en/about-us/news/news/world-water-day-cost-cotton-water-challenged-india/
9. A new concept for sustainable denim production (2013) Available from: http://www.engineerlive.com/content/new-concept-sustainable-denim-production
10. National Geographic (2010) A special issue on "Water: our thirsty world"
11. Pal H, Chatterjee KN, Sharma D (2017) Water footprint of denim industry. In: Sustainability in denim, Elsevier Publications, http://dx.doi.org/10.1016/B978-0-08-102043-2.00005-8
12. Waterprint (2016) (online). Available: http://waterprint.net/jeans.html
13. Luiken A, Bouwhuis G (2015) Recovery and recycling of denim waste. In: Denim; Manufacture, finishing and applications, Woodhead publications, pp 527–540
14. Shaikh AM (2009) Water conservation in textile industry. PTJ, pp 48–51, Available(online): http://www.sswm.info/sites/default/files/reference_attachments/SHAKIH%202009%20Water%20conservation%20in%20the%20textile%20industry.pdf)
15. Uzal N (2015) Effluent treatment in denim and jeans manufacture. In: Denim; Manufacture, finishing and applications, Woodhead publications, pp 541–561
16. Parvathi C, Maruthavanan T, Prakash C (2009) http:// www.indiantextilejournal.com/articles/FAdetails.asp?id=2420
17. Cattoor T (2007) European legislation relating to textile dyeing. In: Environmental aspects of textile dyeing, Woodhead publishing, pp 1–29

18. Sewekow U (1996) How to meet the requirements for eco-textiles. Tex Chem Colorist 28 (1): 21–27
19. Nadiger GS (2001) Azo ban, eco-norms and testing. Indian J Fibre Text Res 26:55–60
20. Intertek (2012) Eco Textile Services (online): downloaded from http://www.intertek.com/uploadedfiles/intertek/divisions/consumer_goods/media/pdfs/services/eco-textiles.pdf
21. Parliamentary Office of Science & Technology, Houses of Parliament. The Environmental, Health and Economic Impacts of Textile Azo Dyes, London. www.parliament.uk/post downloaded from http://docplayer.net/23866015-The-environmental-health-and-economic-impacts-of-textile-azo-dyes.html

Making the Change: The Consumer Adoption of Sustainable Fashion

Alana M. James and Bruce Montgomery

Abstract This chapter explores the recent societal changes leading to a shift in consumer needs and wants from fashion. This aims to offer context to the increased consumption levels in the fashion industry and due to the nature of the necessary supply chain, resulting in consequential negative impacts on social and environmental factors. The fast fashion business model was explored, framing the problems currently challenging an increase in responsible practices in the fashion industry. The adoption of sustainability on behalf of the consumer is crucial, cleaning-up their previous behaviour which focused on quantity over quality. Without adoption, the consumption problem remains ignored and disguised up by aspirational fashion marketing, blinding the consumer during purchasing. The detoxification of both the consumer's purchasing behaviour and the process and practices implemented in the fashion industry is needed. Potential solutions and approaches will be discussed helping to move away from the previously highlighted fast fashion principals, towards a slower, more considered fashion sector.

Keywords Sustainable fashion purchasing · Consumer behaviour change
Transparent business practices

The original version of this chapter was revised: For detailed information please see Erratum. The erratum to this chapter is available at https://doi.org/10.1007/978-981-10-4777-0_4

A.M. James (✉)
School of Textiles and Design, Heriot-Watt University, Galashiels, UK
e-mail: a.james@hw.ac.uk

B. Montgomery
Faculty of Arts, Design and Social Sciences, Northumbria University,
Newcastle upon Tyne, UK
e-mail: Bruce.montgomery@northumbria.ac.uk

© Springer Nature Singapore Pte Ltd. 2017
S.S. Muthu (ed.), *Detox Fashion*, Textile Science and Clothing Technology,
DOI 10.1007/978-981-10-4777-0_3

47

1 Introduction and Background

The past 20 years has also witnessed a gradual growth in society's awareness and concern for ethical and sustainable issues and is said to be due to academic interest, increased media levels and a greater choice of green products available [80]. As a reflection, much has been written both in academia and the media leading to the debate of some key issues within the fashion market sector. In addition to increased levels of ethical awareness, on the contrary there have been a series of dramatic changes within the fashion industry that have impacted both consumers and retailers. The growth of the value sector has spread into the fashion market with the fast fashion movement emerging, which has given consumers a taste for trend-led products, in large quantities at a very cheap price. This has resulted in consumers now preferring large quantities of cheap clothing in preference to a lower number of higher quality items. Possible reasoning behind this change in purchasing responsibility has in the past been suggested, although the rationales have appeared to be quite vague and predictable. Ideas such as price, accessibility and aesthetics have been put forward, yet the depths of consumer behaviour and purchasing hierarchies have been left out of the debate. It has been suggested that consumers harbour feelings of low perceived effectiveness, believing that their change in buying action alone would not make a difference to the wider picture [43], while it has also been suggested that ethical consumerism in general is an overlooked and under-researched area [87].

The growing awareness of ethical issues can also be directly related to pressure group support, media interest and increased corporate responsibility action on part of the retailers [99]. However this change in consumer attitudes cannot be used as an indicator for a change in behaviour. The response that the industry has had in reaction to consumer demand has been labelled as chasing the ethical pound, with funded projects such as Fashioning an Ethical Industry [45] and fashion events such as Esthetica [20]. However despite numerous positive schemes the ethical fashion market remains at just 1% of the overall apparel market [81]. With this in mind, change in terms of both consumer and retailer behaviour in an ethical fashion context is paramount.

The emergence of the ethically aware consumer can be traced back to the 1960s, which has since forced retailers to take a pro-active approach to the provision of ethical produce [25]. This awareness and recognition of retailer action has since escalated with half of the population said to have bought a product or recommended a company due to their ethical credentials [32]. However it is this consumer intended behaviour translating into actual behaviour that in recent years has been identified as a prominent issue in ethical fashion purchasing. Previous research has identified a distinct disparity in what consumers say and do, which has consequently been labelled the intention-behaviour gap [8, 16, 19, 26, 32, 82, 107]. Whilst this has been investigated previously, no definite solution has been identified and consequently ethical research continues to identify such a gap.

The need for fashion to be detoxed has never been more relevant. There has been a dramatic increase in social and environmental non-compliance in the fashion supply chain including that seen in 2013 during the Rana Plaza disaster and more recently, identified slave migrant labour in clothing factories. There is also constant irrevocable damage to the environment caused through the sourcing of man-made fabrics and consequently negative effects such as carbon emissions, all in the pursuit of cheap clothing.

This chapter will explore the key arguments in the consumer adoption of a more sustainable approach to the fashion industry. The chapter will start by providing an overview and background context to the problem space, including how consumers have become accustomed to a large quantity of garments opposed to smaller quantities of more quality pieces. The fast fashion business model as a prominent catalyst to this adoption of increased consumption will be explored in-depth, with examples of social and environmental non-compliance discussed. The problems preventing change will then be explored with consumers and their behaviour being the focus of the discussions. The potential to move to a more responsible and slow-fashion approach will conclude the chapter, looking forward to a more sustainable fashion industry in the future.

2 A Change in the Fashion Market

The generic UK fashion high street has been through a number of recent changes. Extensive research investigating these additional changes was commissioned by the UK Government in 2011 and carried out by industry expert Mary Portas. In recent years, Portas has been active in her study of high street retailers often concentrating on small, privately run stores and British made goods. In her 2011 report she begins to untangle the reasoning behind recent changes on the high street and the potential steps that need to be taken in order to save the traditional high street. These contributing factors include the growth of out of town retail parks draining traffic away from the high street, major supermarkets and malls offering convenient, needs-based retailing and the issue of high streets not keeping up with the changing needs of consumers [86]. The presence of this investigation emphasises not only the identification of changes within consumer shopping habits, but also the need for further research and a strategy for how these issues can be solved.

The rise of supermarkets is another major concern for the well-being of the high street, with every £1 being spent in shops, 50 pence of that being spent on food and groceries in supermarkets [86]. The services that supermarkets are offering is also growing with opticians, tanning centres and travel agents not looking out of place in the average supermarket. It is this element of convenience that the high street is no longer offering to consumers. This is also true in terms of ethical produce, with supermarkets now offering Fairtrade and organic options in a range of goods. Literature relates this to the clothing industry stating that supermarkets have been forced through consumer demand to provide ethical alternatives and question if the high street could go the same way [90]. This idea of consumers controlling the goods a retailer sells, shows that the awareness levels of consumers is growing. It can however be questioned if it is the retail giants that dictate how we shop, or the other way around [50]. However, it is also believed that brands with a strong vision, USP (unique selling point) and ethos are said to be the most successful, not those who extend and adapt their range of supplied goods to fit a market [6].

In the past concern has been placed on the idea that all UK high streets were beginning to look the same. The term monoculture has been coined by researchers [50], while other suggestions have included clone town Britain [86]. This issue suggested that no high street was individual, but played host to the same brands up and down the country. However it could be argued that this repetition is no longer relevant, with the number of town centre stores falling by almost 15,000 between 2000 and 2009 with an estimated further 10,000 losses over the coming years. That equates to approximately one in six shops standing vacant [55]. However this loss is not limited to small and local businesses, the closure of long-standing UK retailer Woolworths in 2009 and BHS in 2016 is a prime example that these changing behaviours can influence even large retailers. Another example of this is the recent announcement from Arcadia boss Sir Philip Green to be reducing the number of stores across the country, whilst he continues to extend the number of Topshop stores abroad [86]. This business decision is a direct reflection of the current UK market and the acknowledgement that business may be more successful elsewhere.

The value market sector has made an increased appearance on the UK high street, with sales reporting a growth of 6%, taking its value up to a huge £8.1 billion. Consumers now prefer to buy quantity over quality, which has consequently had a large impact on the high-end luxury sector of the market. The ultimate advantage the value stores have over the luxury brands is that due to an increased number of range deliveries they can stay more up to date on trends whilst remaining competitive on price. Research shows 36% of all adults now shop at the cheaper end of the market, with Primark being their store of preference [75].

It has been argued that an increased number of offers in-store such as BOGOF (buy one get one free) are seen as an excuse to inflate prices and promote excessive consumption [6]. Recently there has been a growth of the pound shop and the preference value-conscious consumers are paying to this market share in preference to offers being provided by the more traditional store [86]. It is this bargain-hunting hunger that allows consumers to feel they are still able to purchase goods, (often a larger quantity) despite the effect the recession may be having on their financial situation. It has been suggested that the actual experience of purchasing is often the reason a consumer buys goods, with thrill of the buy often being more important than the purchase itself. The increased use of credit is also said to have had an effect on excessive consumption and lack of values. This is due to people no longer having to diligently save in order to be able to afford goods. However it can be argued that the real value clothing sector lies in stores such as TK Maxx, who offer designer goods as a much lower retail price. Thus, purchasing a higher quality garment for often, a fast fashion price [6].

A further factor that must be considered in the exploration of ethical fashion purchasing would be the impact that the recession has had on consumer purchasing behaviour. Consumers are naturally spending less and reverting back to basics rather than consuming products that have ethical or sustainable credentials. It has been suggested that due to the recession the clothing market is not being properly represented and may be in a temporary state of flux due to the impact effecting individuals differently. Acknowledging this however, could have both a positive or negative effect. On one hand consumers could refer to value stores and look for quantity as a substitute for quality, yet on the other hand a lack of finances could encourage a shift to investment purchasing, buying less overall for a higher quality. In addition to this, research shows that women appear to be hooked on fast fashion and that the recession does not appear to be changing their attitudes [75].

It has been well recognised that since the economic downturn in 2001, ethical consideration on the part of the consumer has dropped considerably [107] with more important factors replacing ethical considerations as Arnold [6] reiterates, 'People start from self, spread to family and community and only then to planet, we are programmed to protect ourselves and others close to us'. This is reflected in Maslow's Hierarchy of Needs, which aims to categorize the varied level of needs a human requires in five key stages; *physiological, safety needs, social needs, esteem* status and *self-actualization* (in ascending order). Research shows that a small sector of better educated, well-off society are moving to the top of Maslow's hierarchy into the self actualization sector [98]. This is where consumers have a need to be involved in something outside of themselves, something precious and almost spiritual. When considering Maslow's theory in relation to purchasing behavior, consumers must reach this final category in order to consciously purchase clothing responsibly. However, a further argument can be highlighted, with research suggesting that ethical consumers can be classified as wealthier and more educated than those who do not purchase ethically [105].

This research acknowledges dramatic changes on the UK high streets in recent years, yet leans towards the idea that it is simply the high streets that are not keeping up to date with consumers changing shopping habits. It appears the fashion customer wants convenience when shopping and it is the out-of-town retail shopping centres and supermarkets that appear to be offering this. Free parking, public facilities, warmth and shelter are just some of the convenient advantages that large retail outlets have over the high street. In addition to these, supermarkets also offer value goods at cheap prices whilst being able to be purchased during the weekly shop. In order to compete with these attractive factors, the high street needs to become more diverse and user friendly, offering features such as ethical and organic alternatives. The UK high street ultimately needs to be innovative in their approach to encouraging consumers to return to the traditional town shopping areas.

2.1 Delivery to Market

The change in consumer purchasing behaviour and the need for quantity in preference to quality has consequently resulted in a dramatic change in the traditional fashion cycle. The 2010 WGSN (World Global Sourcing Network) Report [106] states that new seasonal phasing is increasingly on its way and shoppers should be prepared for up to 12 constant-rolling collections per year. This indicates that the traditional fashion cycle is moving from that of a more traditional two seasons per year to a fast fashion business model, which delivers new trends to consumers on a regular basis. This drip-feed of trends throughout a season would naturally encourage further consumption as consumers feel the need to be up to date with trends and peers.

The WGSN report suggests that pressure on designers to produce extra collections will only intensify, with stores looking for newness, exclusives and more commercial lines away from what the catwalk currently offers [106]. In addition to the high street brands, luxury designers are also feeling the pressure for an increased number of collections. The introduction of pre-collections is ultimately encouraging the move towards transitional seasons. Ethically conscious designer Stella McCartney has started working to four seasons per annum, introducing her additional Pure Summer and Winter collections. Professor Wendy Dagworthy, then Head of Fashion and Textiles at Royal Collage of Art defended this change of delivery describing it as just one collection but designed with staggered delivery in mind to meet the need for fresh designs [106].

These highlighted changes in the fashion industry as a whole, continue to reflect a shift in consumer behaviour. The high street, and increasingly high-end designers, are acknowledging the consumer thirst for a continuous stream of trends being leaked into stores. It is this element of adaptability and flexibility that the traditional high street needs to also mirror. Where perhaps consumption can be kept to a minimum reflecting Dagworthy's approach, keeping the size of the collection the same but changing the delivery schedules in order to satisfy the consumer needs and wants from fashion.

2.2   The High Street's Responsible Offering

The high street in recent years has undertaken a number of large changes, with not only the consumer needs changing but also the type of stores available. Portas [86] believes that the high street has not kept up with these changes and as a consequence many shops now stand vacant. This is also true for the facilitation of the ethical market, with ethical food produce being readily available but not fashion garments. Ritch and Schroder [90] also concur with this point, stating that the ethical food and cosmetic market appear to be catered for, but fashion remains lagging behind. On the generic UK high street brands such as Lush and the Body Shop, who offer organic, non-animal tested products are readily available, whilst individual stores and supermarkets facilitate the ethical food market. It is this inconvenience of ethical fashion availability that may be to blame again for the current small market segment. Carrigan and Attalla [25] believe that consumers do not need to be inconvenienced and that ethical fashion products need to be readily available for purchase. This disparity in the availability of food and cosmetics in comparison to clothing has been blamed on the clothing supply chain being more complex and therefore less transparent. It is thought that more details need to be provided, such as country and grade of manufacture for the apparel industry [81], which would only make the auditing process more lengthy and complex. A further argument is the social status that ethical produce currently has, with organic or Fairtrade food thought to be quite fashionable and trendy [90]. This shift to social acceptance of ethical fashion could be the key to raising not only the consumer awareness of ethical issues but also increase the overall market segmentation in the apparel sector.

In recent years high street brands have begun to either house ethical or sustainable brands or produce small in-house ranges that take on an ethical angle. Examples of these would be fair trade fashion pioneer People Tree in Topshop and H&M's production of the recycled fibre range Eco-conscious. Allanna McAspurn, UK manager of Made By (A European not-for-profit organisation with (a mission to make sustainable fashion common practice and improve environmental and social conditions in the fashion industry) believed that such inclusion of ethical ranges on the high street is a positive step forward for the fashion industry. She thinks that they will be beneficial in terms of raising consumer awareness levels and raising the ethical fashion profile. Whilst brands such as People Tree work towards extending their socially aware policy throughout their supply chain, it could be questioned that if H&M's Eco-conscious range claims to affect only the raw material, are they aware of the conditions these garments are produced under? The savvy high street consumer should be asking such questions. Niinimaki [81] raises the issue of value in clothing, and suggests that this can be created when meeting the need of consumers, consequentially extending the lifecycle of garments.

In contrast to additional ranges being added to retailers to highlight their responsible values, companies are now beginning to adopt a more holistic, integrated approach to social and environmental commitments. The movement has seen companies embedding these values into the core of the way they do business, taking these into consideration in their everyday business practices. This adaptation is being adopted by companies such as Marks and Spencer who have underpinned their core values through their Plan A scheme, which functions across all aspects of their business practices. This approach allows for the incorporation of business ethics opposed the an additional side-line, which has the potential to send mixed messages to their customers.

3 The Fast Fashion Business Model

Fast fashion aims to bring catwalk inspired fashion to the masses as quickly and cheaply as possible. This has resulted in a change in fabric and manufacturing quality of clothing, and human compromise in the supply chain [46]. Spanish multinational Zara are said to be a pioneer in the development of the fast fashion movement. However, Zara originally achieved this by using local production and sourcing strategies, allowing them to deliver up to two shipments per week to store, averaging up to 12 times faster than other retailers. This method is said to have set the bar for other retailers to follow. However, where Zara was once changing the industry, the industry is now changing Zara, as other retailers have adopted alternative methods to meet the consumer demands of cheap fashion delivered quickly [103]. Despite Zara initially thinking this was an effective business strategy to take, it was also thought that fast fashion goes against the very premise of the fashion industry and is merely a copy and paste exercise, not creative excellence [12].

The consumer demand for fast fashion has seen a change of consumer attitudes and values moving to a preference of quantity not quality when seeking their next purchase. However this appeal is described as a basic human need to belong to a specific group of society, which consequently encourages over consumption due to the traditional bi-annual changing fashion trends [21]. Young consumers are said to be the most active group of society interested in keeping up with fashion trends and consequently are the biggest consumers of fast fashion [77]. On the other hand, it has also been highlighted that there has been a change in experience when purchasing fast fashion, moving from that of a high-involvement, pleasurable experience to low-involvement and disposable [90]. It is this change in both experience and purchasing behaviour that this research will concentrate on, attempting to outline the rationales behind these changes and providing potential solutions for the reverse of current behaviour.

The current high street leaders in providing fast fashion goods to consumers is Primark, whilst being associated with social supply chain disasters such as Rana Plaza have reported profits for November 2013 up by 44%, with an increase to £514 billion [58]. With this in mind, it has been highlighted that the fast fashion speed to market (fewer pieces, more often) has resulted in retailers collecting a higher net margin of the retail sale value [103]. However, when taking into consideration that a fast fashion item is manufactured to be worn only 10 times [95], it is not surprising that the fast fashion customer needs to continue consuming clothing at a rapid pace. This constant turnover of goods has been labelled the Primark Effect [95], with the fast fashion movement being compared to McDonalds; cheap, fast, mass produced, hassle free and reliant on social or environmental exploitation [90]. Many academics acknowledge this speed of consumption [21, 77, 90, 103, 95], other literature however focuses on change stating that the age of frivolous consumption should come to and end in preference for the right behaviour [96].

The production of fast fashion often requires a social compromise within the supply chain in order to meet demanding lead times [46], with consequences of this being discussed previously at the beginning of the chapter. These range from not only large scale disasters such as the collapse of a building, but also factory working conditions, level of wages paid to workers and the amount of hours worked per week. It is said that the production of fast fashion products has a huge number of costs, the scale and details of which are not yet widely acknowledged. From extensive media coverage the term *sweatshop* is now widely used in relation to the production of garments. This term reflects the often less than satisfactory working conditions which are subject to the 100 million garment workers worldwide [64]. These factories are categorised and specially formulated in order to pass audits set in place by even the most ethical of high street names. An audited 5-star factory for example, maybe the shining representative of several poorer shadow factories where the majority of the real labour takes place [57].

Along with poor working conditions and long working hours, poor pay can also be an issue. On average a garment worker receives only 0.5% on the retail cost of a garment [65]. Putting this amount into perspective, a £2.50 basic coloured t-shirt bought on the high-street would result in the workers receiving just over £0.01 per garment, a shocking statistic that the majority of society are unaware of. Former Chief Executive of Marks & Spencer's Stuart Rose, commented on this in the labour Behind the Label 2009 report; 'How can you sell a t-shirt for £2 and pay the rents and pay the rates and pay the buyer and pay the poor boy or girl who is making a living wage? You Cant' [69]. Although production in such factories may be fast, the quality and efficiency of manufacture is questionable, contributing to the lack of productivity [59].

3.1 Social Non-compliance

Within the fashion industry the fast fashion business model has developed with catwalk inspired looks being available to buy on the high street as soon as four weeks after the designer shows. Fletcher [46] has described this as a process resulting in a change of fabric and manufacturing quality of clothing with social or environmental compromise in the supply chain. The emergence of the fast fashion business model has acted as a catalyst for the fashion supply chain, with a dramatic increase of pressure applied in terms of expected speed and delivery to market. This often has a significant impact on the ethical and sustainable conditions in which the goods are produced resulting in social and/or environmental non-compliance. The manifestation of social non-compliance is varied but as has been discussed can often be prevalent when cheap clothing is produced under poor conditions.

On 24th April 2013, arguably the worst social disaster to occur in the garment production industry happened in Dhaka, Bangladesh [13]. An eight storey commercial building in the region of Savar collapsed, containing several clothing factories, apartments and shops. The garment factories within the complex employed approximately 5000 people [109], producing fast fashion garments for UK high street brands such as Mango, Matalan, Primark and Walmart [79]. During the investigation into why the factory may have collapsed, several factors were discovered leading up to the event that could have prevented the 1129 fatalities and a further 2515 injuries [22]. Historically, the upper four floors of the complex were built without a permit [5], with the architect of the original build commenting that the structure of the building was not strong enough to bear either the weight of the additional floors, nor the weight and vibrations of the machinery required during garment production [17]. This advice however was ignored and the companies continued to use the building for garment manufacture. In addition to this warning, the day prior to the collapse saw building inspectors discover cracks in the infrastructure of the building, requesting immediate evacuation and closure. While this warning prompted the shops and banks within the complex to close, one of the garment factories, Ether-Tex, declared the building safe, threatening workers that a months wages would be docked if they did not show up to work [38]. The collapse of the Rana Plaza complex is just one example of the social compromise that can occur during the production of cheap, mass-market fashion. It is this type of social compromise that prompted this investigation, with the aim of the insights found to contribute to social change and development in the production of fashion garments.

The scale of the Rana Plaza disaster has lead the collapse to be considered as not only the deadliest garment factory accident in history but also the deadliest accidental structure failure in modern human history [14]. As a consequence, it was thought that sales of fast fashion, high street retailers such as Primark would be affected, with the large-scale media coverage having an influence on consumer purchasing choice and behaviour. However this was quite the opposite, Primark annual profits for November 2013 saw a 44% increase to £514 billion, with a revenue increase to £4.3 billion equating to £11.7 million per trading day [58]. Many thought that Rana Plaza would be the turning point of the consumption of cheap fashion, with consumers seeing the effects of the production of such clothing further down the supply chain, however it can be seen in Primark's profits alone, that consumer priorities when purchasing fashion appear to be elsewhere.

Despite the scale of the collapse of Rana Plaza, it is far from the only incident that has occurred in recent years. From late 1990 to the present day, there have been 28 reported incidents in garment factories, with 22 of these having fatalities. During this time, almost 2000 factory workers have lost their lives, with this figure being from only the incidents that have been reported [18]. Other recent disasters include the Tazreen Factory fire and the Spectrum Sweater Factory collapse, accounting collectively for a further 200 worker deaths [22, 70].

On 24th November 2012, a fire broke out at the Tazreen garment factory, on the outskirts of Dhaka, Bangladesh [15]. The fire killed 124 garment workers and injured a further 200 [2]. These fatalities however were again due to breaches in building safety standards, where in this case there was a severe lack of fire exits, with existing exits being very narrow and all leading to the source of the fire on the ground floor rather than to outside the building [4]. Tazreen produced garments for brands such as the Edinburgh Woollen Mill, C&A and Walmart [30], however it was Walmart that flagged this factory in 2011 to have violations and/or conditions which were deemed to be high risk. Highlighted almost a year prior to the fire break out, the incident could again have potentially been avoided [10].

On 11th April 2005 the Spectrum Sweater factory collapsed in Savar, Dhaka, killing 62 garment workers and injuring a further 84. After investigation, it was found that the factory floors that collapsed causing this incident, had been illegally built and therefore did not meet building standards and regulations [70]. It has been incidents such as Rana Plaza and Spectrum that have prompted initiatives in the clothing industry to be developed in hope that this will prevent further occurrences.

It can be highlighted that the three factory disasters discussed have all occurred in Bangladesh, perhaps due to it being the second largest garment manufacturing country in the world [85]. Since the Rana Plaza disaster, over 100 apparel corporations have signed and agreed to the Bangladesh Accord, an independent agreement designed to make all garment factories in Bangladesh, safe work places. UK high street retailers such as M&S, Next and Primark have all agreed to the Accord [1] however it has come under heavy criticism due to the agreement only covering standards of fire and building safety. It could be argued that whilst a building may be safe to work in, the workers may endure other issues of working conditions such as illegal, long hours and wages not meeting minimum wage standards.

This type of agreement has been seen before in the garment production industry with the Asian Floor Wage initiative developed over a number of years and launched in 2009, where trade unions and labour activists across Asia came to an agreement. This was to ensure that garment workers across Asia receive a fair living wage in accordance to the living costs of their specific home country [7]. This initiative was again agreed to by a number of high street fashion retailers, however again this faced criticism due to its restrictive nature of addressing multiple social issues.

3.2 Environmental Non-compliance

In addition to social issues within the fashion supply chain, the environmental impact is also significant. Again this spans the whole of the production process from the sourcing of fabrics to the transportation to retail location. Environmental impact however also continues long into the use phase of the garment with consumers being heavily responsible for the negative impact on the environmental. Care and use of a garment uses extensive amounts of water and chemicals, especially in the washing and drying of products. The disposal however is also prime for potential impact on the environment. Consumer consideration for methods of disposal and care for product longevity can have a great positive change in the life-cycle of a garment which also requires focus in research and literature.

One of the most prominent issues in the fashion industry is water consumption. The use of cotton as a natural fibre consumes large quantities of water not only during irrigation of the crop but also during the dying process. The production of a simple cotton t-shirt for example uses 27,000 l of water through the manufacturing process [108]. This intensive use of water has had a dramatic effect on the effect, especially in cotton growing regions such as Uzbekistan. Over recent decades this area has suffered profoundly with Aral Sea shrinking to only 10% of its original size between 1960 and 2007. Once one of the world's four largest lakes at a size of 68,000 km^2, this split into four lakes but in continuous decline has no been renamed as the Aralkum Desert. This is just another example of the hugely negative impact the fashion industry has on both social and environmental aspects of society.

4 The Detox of Consumer Purchasing Behaviour

While the problems of responsible fashion purchasing have been discussed, the detox of such behaviour must next be addressed. In order to do this, social and environmental responsibility must be taken back to basics with the consideration of moral values and ethics. Aristotle described human beings as rational animals, implying that responsibility begins with reasoning using techniques of logic, science and analysis [9].

Defining ethical as an understandable term has proven difficult for researchers and academics alike. There is currently no industry standard or working definition for the term and consequently is often a misunderstood and confused area. It has been said that the definition has become an issue with factors often being subjective or situational [19]. The lack of precision in defining this area has resulted in an array of inter-related terminology being used [100]. It has been utilised to cover a range of activities in the clothing industry including; worker rights, transport, chemicals used and the actual processing of the garment (dying, finishing etc.). However this raises an argument that if a garment is ethically compliant in terms of raw material, for example, but then is produced in a factory that does not meet regulation, is the garment ethical or not ethical? This argument could also be reflected in the brand of Fairtrade. This certification currently only refers to the raw material, in the case of the clothing industry this can only be applied to cotton. It has therefore been suggested that due to the whole of the supply chain not being covered the application of a certification can be contradictory and misleading for the consumer [45].

To refer to the origins of the term ethical, it derives from the meaning arising from character, the Greek ethikos or ethike and the Latin moralis. They also carry the connotation of arising from habit or custom [9]. These definitions however rely on the subconscious of the individuals being aware of what may be wrong or right behaviour. This appears again to be a very hard area to define and could be described as far too subjective to be relied upon. Taking quite a realistic viewpoint when discussing ethics, it has been suggested that the term ethical is far too broad in its definition, too loose in its operations and too moralistic in its stance [36]. They continue to reiterate an earlier point, leading them to the conclusion that ethical consumption is therefore a myth. This argument again raises the issue of ethical awareness levels amongst consumers. The individuals perception and understanding of the term could also be an issue. It is acknowledged that consumers do not currently have enough information and understanding of the term ethical to make a fully informed purchasing decision [90]. Another practical line of reasoning would be the idea of a moral relativist, who believes that all people do not hold or obey by the same morals and ethics during their day-to-day lives [9]. Again this reflects an earlier point, into the different ways individuals will choose to use their own ethical values and ultimately how these attitudes are translated into actions. This research aimed to take a more realistic approach to this line of enquiry, trying to acknowledge that individual consumers are different in their attitudes and behaviour and that large grouped assumptions are impractical.

4.1 The Responsible Fashion Market

The ethical clothing market remains at just 1% of the overall apparel market [81], however it can be argued if this 1% market share meet the needs of ethically aware consumers. There has been much literature published regarding the high street not offering adequate amount of ethical alternatives or options [25, 90]. The low percentage of ethical market segmentation in the apparel market has also been blamed on ethical produce often costing more, or in the case of fair trade, a premium being added. Although the logical argument would be, if value sector garments are produced in less than satisfactory conditions, then in comparison ethical produce will cost more. However it is well recognised in previous research that consumers would make more ethical choices if they did not have to pay more for these options [19]. In addition to a price reduction consumers would also like further information and more choice in the range of clothing available [90]. Devinney et al. [36] however argue that the most successful ethical products, are those that are not only ethically compliant but those that also fulfil a market niche. This approach reflects that being taken by this investigation, acknowledging that to develop the ethical market, a more mainstream attitude needs to be taken. The idea of integration is also highlighted by Niinimaki [81] where it is questioned why there appears to be two separate markets, ethical and non ethical, and why this gap remains so.

4.2 The Responsible Consumer

Devinney et al. [36] believe that the term ethical consumer needs to be approached with caution as there are many influential factors that can effect the decision making process. Wehrmeyer [105] goes as far as narrowing down the ethical consumer profile said to be educated, urban, AB socio-economic and often married with double incomes. Although this research acknowledges Wehrmeyer's point of view, this study recognises that this level of clarity and precision cannot be relied upon. Through the research process and the literature search it can be identified that there are many contributing elements to the consumer and their attitudes and behaviours, and that such a linear approach to the ethical consumer profile cannot be taken.

It is to be acknowledged that every consumer is individual regardless of their background and interests, and therefore the key is to identity their specific needs [12]. However previous research has tended to group consumers together in creating a typology [25, 31, 32, 74, 77, 100], where realistically there is little consistency in demographics and consequential behaviour [36]. These typologies aim to categorise consumers into generalised groups ranging from the non-ethical to the super-ethical. For example, Clouder and Harrison [31] divide consumers into three key groups; distancing, integrated and rationalising.

It is believed that the lack of connectivity and therefore compassion towards the social factors in the garment supply chain can often lead to non-ethical purchasing. Carrigan and Attalla [25] believe that the consumer importance of self continues to emerge, where if unethical behaviour affected them personally, they may care more. Perceived consumer effectiveness (PCE) has also been said to have an influential impact on ethical purchasing. This is the extent to which a consumer feels that their individual contribution will make a difference [43]. This relates directly back to consumers wanting to feel like they are part of a group or tribe [21], which is where a consumer typology has the potential to have a positive influence on ethical consumer behaviour.

It has been found that there are more conscious consumers than previously thought [100] with 57% of consumers claiming that they would not shop somewhere if the store or brand had affiliations with child labour [25]. However it can be argued that this expression of ethical concern does not reflect the current statistics of the 1% ethical market segment [81]. The more realistic viewpoint perhaps comes from Devinney et al. [36] where the actual existence of the ethical consumer is questioned, stating that the ethical consumer is a myth due to it representing a role model that does not currently exist. This ideology is recognised in this research, acknowledging that consumer behaviour could be described as fickle and very unpredictable. In preference to the term ethical consumer, it has been suggested that consumer social responsibility is used, which can be defined as the deliberate choice to make certain consumption choices based on personal and ethical beliefs. This approach describes ethical consumption in a more real-time manner, suggesting it to not yet be habitual but however possible depending on the individual.

Plato and Aristotle stated that life should be based on a series of good actions and that we should strive to be good and virtuous in those actions. This can be directly applied in the process of responsible consumption, implying that individuals should take a more in-depth view of, in this case the clothing supply chain, prior to making a purchasing decision. However it is said that well-informed consumers are challenging retailers in their ethics [99], taking a much more hands-on approach to demanding the availability of ethical goods. On the other hand, consumers are said to care more about the colour of a trainer, for example, than the conditions in which it had been produced [36]. The exact typology of the consumer cannot be pinpointed, whereas the behaviour of an individual also cannot be categorised. Where a consumer may purchase in an ethical manner one day, the next day the behaviour may be completely different. There are too many influential factors intervening in the consumers purchasing process for future or habitual behaviour able to be predicted.

4.3 Consumer Awareness and Knowledge

A lack of ethical behaviour is said to often be due to low awareness levels and an overall lack of knowledge [43]. However, 52% of consumers in the UK admit to be ethically aware but are currently not actively purchasing ethically [107]. This relationship between the knowledge/awareness levels and actual behaviour is an area that this research explored further as a well informed consumer is said to be the key in understanding the ethical decision making process [80].

An underlying argument within the debate of ethical awareness levels is the idea of consumers having enough knowledge (or not) in order to make a well-informed decision. Whilst Ritch and Schroder [90] believe that a fully informed consumer is unattainable, it is also thought that growing levels of ethical awareness is due to academic interest, increased media levels and a greater choice of ethical products [80]. However, researchers believe that consumers think more ethically than they actually do. This is said to be due to weak research methods being used, leading to inflated measure of intentions [26]. Ellen [43] reiterates this point, as consumers are not as knowledgeable as originally thought, and not aware enough to make an informed decision.

As previously mentioned, previous research has seen many consumer typologies developed, however these typologies can also be developed as an indication of a consumer's awareness and knowledge levels of ethical issues in the fashion industry [25, 31, 32, 74, 77, 100]. These provide categories or levels where consumers can be grouped together in terms of their ethical knowledge and behaviour. Example typology categories include; not noticed an issue, aware but not greatly concerned, aware and concerned but have not taken action, concerned and intended to take action, concerned and taken minor action and concerned and taken major action [52].

Whilst methods of better informing consumers of social and environmental issues, have in the past been criticised due to the true underlying message being hidden in an attempt to make ethical issues more popular [36], a direct link to the amount of knowledge a consumers has and the method used to inform has been found.

Despite levels of knowledge and ethical awareness, it comes down to the individual to use this information to inform their decision making process and purchasing behaviour. The 2010 Co-operative Ethical Consumerism report states that consumers are highly aware of ethical issues and are ready to put their money where their mouths are [32]. However it is thought that the awareness of an issue will only come if consumers think it is sufficiently important compared to other imperatives in their lives [51].

Organisations such as Traidcraft and Café Direct have been persistent in their delivery of both products and information in past years, with a large proportion of consumers now being aware of such brands. It is this persistence and consequently awareness levels that have changed consumer's issue focus from environmental to social [99]. McAspurn from Made-By believes this shift in awareness focus to be true, with more consumers now having a firm ethical awareness in preference to just green environmental knowledge. This research has concentrated on consumer awareness levels of ethical issues, working to the definition and conceptual framework formulated through the study. Whilst a large proportion of consumers do not have a knowledge level that would account as an informed decision, the use of this knowledge remains subjective and individual to each consumer. Innovative and forward-thinking methods need to be adopted to inform consumers, with these methods relating directly to the knowledge and how this can be applied to future behaviour.

4.4 Responsible Purchasing Behaviour

During the purchasing process there are many opportunities for the consumer to be influenced through a number of different channels. These influential factors range from advertising and price promotions, to the weather and even the mood of the individual. All these intervening elements have the chance to change and persuade the consumer to alter their intended behaviour, differing from what they had initially set out to do. In addition to situational factors there are also fixed factors that cannot be altered by chance, but may play a huge role in effecting the consumer's final purchasing decision. Examples of these factors are price, availability, size and design, all of which are out of control by the consumer yet can affect them in the same way as the situational factors.

The recession has had a significant impact on consumers and their purchasing behaviour, with a large proportion of consumers now favouring the value end of the fashion market [90]. The Financial Times highlights this comparison of price to other influential aspects; 'the decision to buy an ethical brand over a conventional alternative is also influenced by a number of other factors including; brand awareness, the importance of other product criteria, the extent to which buying and ethical product implies an inconvenience or product compromise, if at all, and of course price' [35]. Cowe and Williams [32] also question this trade-off of ethics for price asking are ethics overwhelmed by value for money, as price dominates most shopping decisions. Both these arguments presume that people are forsaking ethical choices for a more value product and whilst consumers shop on a budget, ethical purchasing cannot be achieved. Carrigan and Attalla [25] believe that the most important purchasing criteria is price, value, quality and brand familiarity. The fact that ethical is not one of the purchasing priorities indicates that consumers will look for all or some of these factors in a product before ethics is even considered. In addition to socio-economic positioning, it is thought that culture also plays a role within the purchasing hierarchy [16].

A further issue that is often said to impede further responsible behaviour is the lack of choice and availability [6, 19, 25, 81]. As mentioned earlier, whilst ethical food and cosmetics are readily available on the UK high street, clothing remains virtually unattainable. Niinimaki [81] believes that the lack of trend-led clothing is responsible for the limited consumer interest in social and enviromental issues. This idea of compromise arises again, as if the consumer is wanting to purchase ethically they have to renounce trend led fashion for that which is ethically compliant. Arnold [6] claims that the fashion market needs to combine ethical values with key trend aesthetics in order to be successful. However a more realistic view of this compromise situation is to acknowledge that for the majority of consumers ethics is not a high priority when purchasing clothing. It has been widely recognised that ethics adds value to clothing, however it is not the sole choice for consumers during the purchasing process.

In addition to the issues previously discussed, scepticism of product labelling on the part of the consumer could also be an influential factor [19]. As mentioned earlier, there is existing doubt surrounding the effectiveness of an ethical label in terms of delivering an overdose of information to the consumer. This cynicism has also been said to extend towards the brand or retailer behind the ethical claims, with Cowe and Williams [32] believing that corporate cynicism, consumer disillusionment and the disinterested generation of consumers could kill off the ethical movement. Whilst negativity towards labelling and fashion retailers is acknowledged, to influence ethical fashion purchasing, this research worked to explore further the idea of combining ethical values with mainstream trend led fashion. This also refers back to the idea of there being two separate markets opposed to one inclusive ethical industry [81]. However in order to move towards a more mainstream ethical industry, further research and enquiry must first take place

Respnsible purchasing is often related to the trade off or compromise of factors such as price, style of choice in order to be able to purchase ethical goods. The term flexibility has been applied to ethical purchasing, as the need for balancing ethical ethics and values with the practicality of everyday life is needed [100]. It is when this balance of ethics and everyday practicalities cannot be balanced that the consumer may enforce justification strategies. This is almost a rationalisation with their own ethical values as they consciously carry out what they know to be unethical behaviour. This has been described as consumers justifying decisions through the attachment of logic and meaning [8]. This theory does however again rely on the consumer being informed prior to this strategy being implemented. Other justification strategies includes consumer purchasing goods produced in the UK or second hand mass-produced items as a substitute for actual ethical purchasing. These techniques have been referred to as neutralisation, where consumers justify or dilute their unethical behaviour through strategies, whilst denying all responsibility of negative, consequential effects [28].

This leads to the argument of consumers often feeling no connection or compassion with the social impacts on the supply chain. This can therefore again lead to the justification of unethical behaviour as consumers are not aware (or ignore) the negative implications their purchasing behaviour can lead to. Niinimaki [81] reiterates this point claiming that people subconsciously make decisions benefitting either their individual or collective needs. This lack of connectivity also has a reverse argument where consumers feel that their contribution towards ethical behaviour will have little to no positive effect. Ellen [43] coined the term perceived consumer effectiveness (PCE), which describes the extent to which a consumer's contribution is thought to make a difference. This is related directly to low perceived consumer effectiveness (LPCE) describing the degree to which a person feels they have control over their behaviour. When a consumer feels they have little control or effect in their behaviour, it has been proven to reduce behavioural intentions and consequently actual behaviour. Reassurance strategies of consumer effectiveness could be one approach that could overcome this issue.

Throughout this chapter, the notion of a consumer purchasing criteria has become a reoccurring discussion. This is the idea that a consumer has a hierarchy or priority factors that they specifically look for prior to purchasing goods. Ritch and Schroder [90] highlight the fact that it is not only in responsible fashion purchasing that this complexity of conflicting values within the purchasing criteria occurs. This relates to Maslow's hierarchy of needs as it reflects the gradual increased needs from (in this case) clothing and how this can relate to the various stages of the purchasing criteria. Freestone and McGoldrick [52] believe that UK and USA consumers are beginning to show signs of reaching the self actualisation stage that Maslow describes, indicating that they are beginning to consider ethics in their purchasing behaviour. To use a more specific example, it has been indicated that humans require three key elements from fashion; psychological, to be seen as fashionable, physical, body shape or functional purpose and externally elicited work or occasion dress code [72]. This is loosely modelled considering practical elements that consumers need from clothing, without considering specifics in relation to context or the individual at hand.

Whilst working within a purchasing criteria a series of trade offs occur, deciding what factors are compromised for others [52]. This has been described as a decisional balance scale, where trade offs between anticipated gains and losses associated with behaviour are decided [24]. However due to this purchasing criteria being very personal and subjective it would be very difficult to categorise or model such a detailed process. Szmigin et al. [100] reflect this individual approach recognizing that a flexible decision making process needs to be utilised, treating each case as individual whilst considering the contextual and social factors effecting different situations. Recognition of an individual and personal approach to the understanding of the purchasing criteria is widely acknowledged during this research, allowing for a more flexible and adaptable attitude towards the data collection series.

The purchasing process has been described as a three stage process; input, where factors that effect the purchasing decision are considered, process, where the need recognition of a product is recognised, competitor analysis carried out and a decision takes place and output, where the purchase of an actual product and post-purchase evaluations are conducted [94]. However this process has also been considered in terms of an ethical decision process where it moves to a four stage process; recognition, application of ethical judgement, putting ethical actions before that of others and finally, ethical action [89]. This additional stage in the purchasing process is that of ethical consideration, where a moral or value element from existing attitudes may infiltrate the decision of the consumer and effect the final product selection made. Whilst this model clearly categorises a time when ethical considerations should take place, it is also thought that ethical consideration can take place at any point of the product life cycle. For example, ethical contributions can also occur during the product disposal in terms of recycling and/or appropriate disposal. This arguably could also be perceived as ethical behaviour as a moral attitude needs to be implemented in order to achieve the behaviour. Newholm and Shaw [80] consider however a different fourth stage of the process; perceive needs, gather information, the utilisation of their perceptions of the social context and finally develop behavioural intentions. The majority of the decision making process, in their opinion, occurs prior to the development of a behavioural intention, which is not considered by the previous models of the purchasing process discussed.

Whilst ethical purchasing behaviour can be described as the expression of an individual's judgement in their decision process, this can also be expressed in the choice to not purchase a specific product, due to disagreeing with a company's ethics [97]. This way of thinking can also be implemented in purchasing behaviour models, where the majority of models demonstrate ideal ethical behaviour in preference to unethical behaviour. However, this again reflects the point of Devinney et al. [36] where they argue that the ethical consumer is merely a myth due to the model ethical behaviour used to describe such a consumer does not actually exist. This approach could be applied to the ethical purchasing process and perhaps a model of seemingly perfect ethical behaviour also does not exist, and that behaviour can be a much more unpredictable and irregular process.

The majority of purchasing behaviour models are heavily influenced by The Theory of Planned Behaviour [3], where it is thought that behaviour as a theory of attitudes and that behavioural relationships seek to provide an explanation, linking attitudes with subjective norms, perceived behavioural control, intentions and behaviour in a fixed casual sequence. However researchers have since adapted and changed this theory due to results from further research carried out. For example, further areas for consideration includes implementation intention, situational context and actual behaviour control. It is thought that the utilisation of these additional elements can not only be used to demonstrate the purchasing process but also be used to encourage consumers to purchase in a more ethical manner.

Recent research exploring consumer ethical attitudes and behaviour, has lead to the identification of a distinct disparity in the individual's purchasing intentions and the translation of these intentions into actual behaviour [8, 16, 19, 26, 32, 82, 107]. This is more commonly known as the intention behaviour gap, however it has also been known as the 30:3 syndrome [32]. The numeric reference indicates that 30% of consumers claim to have ethical purchasing intentions, whilst only 3% of consumers actually purchase ethically [19]. It has been said that the ethical claims of consumers by far outreaches their actual behaviour, which may be due to social desirability [107]. This idea of consumers providing socially acceptable answers during research has come up as a reoccurring issue in ethical purchasing research, with methodologies being explored in order to avoid this [8, 41]. 52% of the British public claim to be concerned with ethical issues but admit to not following these concerns up with appropriate purchasing behaviour [107]. These statistics confirm that an intention-behaviour gap exists, yet the acknowledgement of this disparity is not enough, the reasoning behind this could indicate how intentions could be translated into behaviour. However on the contrary, the intention behaviour theory needs to be considered, where consumers may go shopping with the intention of not purchasing ethically, but end up purchasing a product that is ethical. In literature this has previously been unconsidered and could again be an area for future research, providing potential insights into the existing intention-behaviour gap.

The reasoning behind the intention-behaviour gap has been the focus of recent research in the area of ethical purchasing. Carrigan and Atalla [25] believe that ethical behaviour is not part of an individual's purchase decision criteria, whilst thinks that there is not enough information provided and therefore the consumer cannot make informed decisions. On the other hand, the depth and methods of data collected is blamed, with Preez [87] thinking that the intention behaviour gap is an overlooked and under researched area and Auger and Deviney [8] blaming the research methods and strategies undertaken. However another angle within the intention-behaviour gap debate reflects that it is the fault of the retailers, with believing that the high street provides a lack of ethically friendly products and Niinimaki [81] thinking that ethical clothing is not trend focused enough. However to take a much more human centred approach, it has been suggested that the lack of ethical obligation is responsible for the intention behaviour gap [82]. This is due to ethical obligation requiring a more emotive and connected approach on the part of the consumer/ product relationship. However this research recognises that there is no one focus of blame for the intention behaviour gap and it is a number of factors, along with a shift in consumer and retailer behaviour that ultimately contributes towards the identified disparity.

A further emerging debate within the area of attitudes and behaviours is the reliability of using intentions to indicate future behaviour [83]. This argument could potentially flaw the theory of the intention behaviour gap, again reflecting the point of Carrington et al. [26], where inflated intentions are considered. While it is thought that intentions do not indicate behaviour, it is thought that beliefs determine attitudes, attitudes lead to intentions and intentions inform behaviour. The use of inform during this theory is crucial, reflecting a far more subjective approach to the intention-behaviour gap debate. Whilst the previous point highlights the need for further understanding into the relationship between consumer attitudes and behaviours, it is also thought that consumer rationales may also bridge the gap between beliefs and behaviours [16]. However there remains researchers who believe that intention's directly indicate future behaviour. For example, Ozcalgar-Toulouse et al. [82] believe that behaviour is a direct function of an individuals intention, whilst intentions are a function of attitude and that attitudes are a direct representation of an individual's beliefs.

5 Challenges of Sustainable Behaviour Adoption

The challenges facing the consumer adoption of sustainable behaviour have been identified during a period dominated by fast fashion production and retail sales. The adoption of sustainability on behalf of the consumer is crucial, without adoption, the consumption problem remains ignored and disguised by aspirational fashion marketing, which blinds the consumer during purchasing. For companies to apply the necessary changes that will aid the supply chain process a concerted and consistent approach will be required.

The Brundtland commission report referred to sustainability as meeting the needs of the present but at the same time allowing future generations to meet their own needs [104]. This statement can be applied to the challenges of sustainable behaviour with the cleaning up of previous behaviour needing to be administered in a way that does not inhibit fashion development or responsible consumer purchasing. The idea of sustainability is rooted in stakeholder theory ([42]) that emphasises that a firm should pay attention and cater to the interests of both primary stakeholders, who are essential to the operation of the business (i.e., customers, employees, and investors), and secondary stakeholders, who can influence the firm's business operation indirectly (i.e., community and the natural environment).

In investigating the challenges of sustainable behaviour adoption a common thread brought to the forefront is that of transparency within the supply chain and how to allow companies to pursue their own production agendas while providing the consumer with a clear insight into demonstrating best practice. It is the consumers who must be better informed if the cycle is to bare any sort of sustainable pattern. Many managers are feeling the pressure to respond to and act on social and environmental needs in their operations ([42]). As such, a growing number of firms have embedded sustainability in their business models, anticipating positive consequences by doing so. Some firms see it as an important way to protect or strengthen their reputation and brand equity, while others take a more proactive approach by formulating sustainability policies, managing their carbon footprint, incorporating sustainability metrics into their operations, or selecting suppliers based on meeting sustainability criteria. This pressure is positive if it encourages action but not if it is done without belief as this only results in poorly implemented strategies and no attention to detail of positive results. There are various walls and barriers to cross within the process of sustainable behaviour adoption although none that cannot be tackled if there were a complete willingness from all parties involved in the process. These barriers can be identified as price, transparency, consumption, social labelling, and adoption itself. Additional barriers which continue to discourage consumers from further engagement with transparent fashion business models include the availability of ethical and sustainable products produced through reputable supply chains and better access to a greater range of sustainable products.

5.1 Implications of Price

One of the first issues to tackle is the perception of a higher price for ethical and sustainable products. Ethical products and the managing of transparent sustainable supply chains comes at a cost. Devinney et al. [37] stated that a willingness to pay is restricted to a minority of consumers and that the majority of shoppers couldn't care less about an organisation's CSR strategies if it means paying a higher price for a product. While good practice does come at a higher cost particularly with paying proper wages and introducing good working conditions in factories, labour research by the ethical fashion brand People Tree. This showed a conscious influence starting to emerge where in a survey of consumers by 74% agreed to pay extra if it mean a more ethical and sustainable supply chain and end product.

5.2   Trust and Transparency

The new relationship between the consumer, manufacturer and brand, based on the need to create a new form of 'trust' is only possible with a new sense of transparency. The definition of transparency is referred to as the disclosure of information which highlights a requirement for further communication between supplier's brands, retailers and consumers [76]. Without transparent communication, consumers generally have relatively little knowledge of the fashion supply chain and how the clothing they are buying has been manufactured or where for that matter. Surprisingly, even garment labelling has scope for companies finding loop holes in labelling regulations. The retailer supplier relationship is one of supply and demand based on volumes and price negotiation. This relationship effectively gives the retailer control to dominate the supplier who does not deliver on time with heavy penalties. It is based on a concept that the customer will only pay a certain price for a particular type of garment. The retailers are with the exception of the independent retail market predominantly large-scale operations that wield considerable power throughout the industry. This is why they are key to the process as this power could be and should be used to make the supply chain more transparent and change the purchasing process so that it encourages social responsible behaviour. It is too easy for retailers to create distance between the supplier and the consumer to the degree that the consumer purchases the product with very little knowledge about the working conditions in which the majority of fashion is made throughout the globe. Whilst garment origin is specified on a tiny label on the majority of garments this does not actually indicate which factory has made the garment. Therefore, the consumers and garment workers are worlds apart, so the consumer has no sense of responsibility for the working conditions endured in the garment making process. Even the buyers themselves do not always see inside every factory they are working with in the supply chain. The research indicates that the barrier, a lack of seeing for themselves and lack of understanding is at the core of the problem for the consumer to purchase fashion clothing in a responsible manner.

5.3   Consumption

The consumption problem remains ignored and disguised up by aspirational fashion marketing, blinding and heavily influencing the consumer during purchasing. In the present climate of increased social and economic instability and profound changes in the global political landscape, it can be questioned what these sustainable developments will mean for business. Will companies feel there are more pressing issues and in order to counter this? And what are the key issues shaping the sustainability agenda that's leaders should have on their agendas? One continues to ask will profit and short term survival continue to be the main priorities. The stated fashion must come clean as if this was enough to spark the industry into a complete overhaul, the point of a continued lack of transparency was made. Fletcher [48] argues that sustainability occurs when social, economic and ecological factors are satisfied equally, and imbalance is less apparent. Our desire for speed in creating volumes of fashion garments is sometimes unnecessary. Short lead times of standard mass production ensure low costs through economies of scale [47]. Where consumer demand for fast fashion garments exists, supply chains will develop to increase fast consumption [61], ignoring any ethical, moral or sustainable considerations. Fast fashion retailers continue to break down the barrier between the designer catwalk and the retail high street, making versions of the latest trends styles at low cost and affordable to the average consumer. The consumer continues to feel spoilt by choice and so cares less about knowing how or where the garments have come from. This process of shifting product quickly means producing a lot at a low price which puts pressure on the supplier. A process the consumer seems little aware of until a factory disaster like that seen at Rana Plaza occurs. Sustainability should also be embedded in the research, design, production planning, finance, accounting, HR, marketing, PR and communications processes. Luxury brands are through their desire for quality and craftsmanship ahead on producing slow fashion but as they become larger and more powerful super brands there is a great danger that they reduce the attention to detail in the supply chain that previously distinguished them from the fast fashion of the high street. Another factor where research has been done but the industry continues to ignore it is within social labelling. As consumer awareness and concern about production conditions slowly increases, and forecasters point to an as yet untapped market potential for ethically-produced goods such as apparel, so the search continues for robust means by which socially conscious fashion consumers, and indeed other stakeholders, can be reassured about the labour practices behind the garments they buy [71]. The development of social labelling systems is complex in supply chains particularly where licensing, and monitoring by third parties are involved. Social labelling refers to attempts to assure consumers and potential business partners that the labour practices connected with the production of a good or the delivery of a service meet a range of internationally agreed employment standards, generally those based on ILO conventions [71].

Meeting employment standards comes at a price and resistance has been high among high street fast fashion brands. There are good examples of market based initiatives and trade bodies such as- Fairtrade, Made in Green, and Made by which can claim to have had an impact on consumer behaviour through persistent marketing and the use of social labelling. Dahl [34] highlights the question, *do we know what we are buying*? Blind purchasing is common as the consumer becomes engaged in the marketing and aspirational process of retail. This emphasises the need for consumer awareness. The point of consumer awareness is at the point of demand and needs to be before production if it is to have any influence of the end result. To be aware once we have already consumed is essentially too late and partially reflective of the current position.

5.4   The Role of the Consumers

McNeil and Moore refer to three types of consumer 'self' consumers, concerned with hedonistic needs, 'social' consumers, concerned with social image and 'sacrifice' consumers who strive to reduce their impact on the world. This implies that two of these groups at least need particular attention if the new slow fashion business model of providing the consumer with a more transparent view of the fashion supply chain is to succeed. McNeil and Moore go on to highlight that these different consumer groups look at fast fashion in different ways and therefore the implications for creating marketing strategies about fashion that they would engage with need to be significantly different. It is apparent through the medium of social media to make consumers far more aware than ever before and to shift attitudes. Reillyn and Hyman [88] describe that their research indicated green firms are far more active in using social media to highlight strategies and discuss issues than non-green firms who shy away. Another study by Kim and Ko [63] showed that there is a link between social media and luxury brand consumer equity and that this allows luxury brands to understand better their consumers purchasing intentions. If this is the case, then social media has a key role to play in breaking down the barriers mentioned in this section.

5.5   Adoption of Sustainable Attitudes

In order to create a new business model one needs to consider present models that are identifying consumer choice. Consumers can be persuaded to make choices through seeing benefits illustrated through rational choice model [60]. This works to some degree provided that the consumers have sufficient information which supports the view of transparency at all stages of the chain hence the need for a new type of business model to address this. Lifestyle choice is another area that has been under analysis recently, including a DEFRA (Department for Environment, Food and Rural Affairs) project which explored sustainable lifestyles. It developed the 4E's model, highlighting that there is no one single approach to further engage consumers in more sustainable behaviour. The research states consumer behaviour can be influenced by the 4E's; *enable, engage, encourage* and *exemplify* which shows different approaches to identifying consumer purchasing behaviour. In contrast to the policing of supply chains this is a theory which proposes that supply chains can be very positive for developing communities. Developing supply chains in poor and vulnerable communities but not as a destructive force can have positive outcomes [53]. If conducted in a controlled and ethical manner the income generated for these communities can be vital for their survival and development.

The availability of ethical and sustainable products produced through reputable supply chains and a better access to a greater range of sustainable products needs consideration. First it is worth considering that production of new sustainable products needs to be met if we are to move forward. New machinery, adaption of production methods increased overheads all need to be tackled to accomplish this goal. This needs to be considered both domestically and internationally as although not all countries are thinking in the same level of sustainable development the issues are global. The individual consumer also needs to be considered as well as the collective consumer. The influence of our peers can be fundamental in our purchasing choices. In the light of consumer purchasing and consumer awareness the consumer should really be asking themselves *do we really need it*? This is a straight forward purchasing choice that is often completely ignored and does not actually involve any other factor apart from possibly desirability or aspiration. It is evident to us that our wardrobes are already saturated [101] and the possibility for mending and recycling needs to be considered further.

When this individual attitude is transferred to corporate and company levels the scale of tackling an adoption of more sustainable behaviour is enormous. Once we assume these are in place then better access is required through communication that engages rather than dictates. This communication through engagement is a strategy that has been pursued by trade bodies such as the Ethical Fashion Forum, who have been promoting sustainable fashion brands and good ethical practice for years but are still seen as a niche market. Esthetica at London fashion week is now the largest exhibiting group area within the full static exhibition yet it is the glamour of the catwalk presentations that capture the media, buyers and press. Awards like the recently introduced Kering award for sustainable fashion from the luxury group, often have a strong impact by providing news, credibility and action. Some of these catwalk designer brands such as Gucci and Stella McCartney are making great efforts to improve their ethical and sustainable strategy and practice but there is little consistency. The highlighting of pollution, carbon footprint, health, over consumption, treatment of workers and low wages, are becoming more and more the concern of the consumer but not to the degree that sees any major sway in the overall fashion purchasing. The consumer shows concern but then continues to follow traditional spending habits and retail therapy. The Sustainable Apparel Coalition which has Adidas, American Apparel, Burberry and Gap amongst its global membership has strong research and initiatives such as the HIGG index that link the retailer, manufacturer and brand but surprisingly appear to leave out the consumer. 'It seems we address ethical or ecological issues in every other part of our lives except in fashion' [68]. Obvious measures such as taxes, or mandatory labelling and symbols such as in the food industry would be worthwhile to consider. Taxes should be staggered to link with consumption levels so as not to pick on disadvantaged consumer groups.

5.6 Marketing and e-commerce

It is worth noting that the large e-commerce fashion sites rarely show any information that indicates the level of good practice of the supply chain involved. The country of origin is hidden with the garment, with the brand name as the only point of reference for the consumer. If in some case the brand has made a determined effort to highlight how and by whom the garment was made. Then the brand name can act as a purchasing guide but this a very limited approach. Some new independent brands such as Zady (a lifestyle destination for conscious consumers), have on their website made a specific point of showing the whole supply chain and as much of the creation process as possible. This has been commended and could act as a way forward provided these brands manage to maintain enough profit to be sustainable themselves. The site shows where and how the garments are made, the type of stitching on the garment and even highlighting levels of water consumption and carbon footprint. This is incorporated into the website via a combination of the ethical digital magazine and e-commerce shop. The mission statement and manifesto is clear and a series of consumer based articles and links engage the consumer. A simple philosophy of 'keep it, love it, use it' is aimed directly at the consumer.

To summarise there are pockets of very good transparent practice within the fashion supply chain and sustainable initiatives are increasing but not at a rate or in a sufficient critical mass to administer real change. It is still short far too easy to hide behind green wash marketing and appear to be abiding with sustainable practice when a price profit strategy is still driving the business. So to coin a well-known fashion phrase, *if green is finally the new black* then the number of fashion brands that can be classed as ethical and sustainable should be endless. Fabric, style, garment cut, trend and fit should not deter from fair trade, good working conditions and employee working rights. The process should be seamless. If a slow fashion approach is implemented, then both a large percentage of the high street and luxury fashion industry should be able to modify their production practices to have a collective impact that allows for a more transparent supply chain that they would have no problem in allowing the consumer to engage with as there should be no bad practice to hide. The next section will look at what the new business model could look like.

6 Moving to a Slow Fashion Approach

The three P's, *people process* and *profit*, could be seen cynically as supplying vast volumes to people through a fast process at a high profit. This has exaggerated a climate of low wages, poor working conditions, excess consumption and a consumer that is starting to subconsciously devalue clothing due to an over-abundance of low cost clothing being thrown away. Thanks to the rise of fast fashion, the fashion industry is viewed as a main contributor to the above mentioned problems.

This attitude was captured in Andrew Morgan's film, The True Cost [102] during which the consumer body initially appeared concerned, but then due a lack of continued transparency the consumer falls back into a more non-confrontational mood. It should not be acceptable that retailers can offer products that are not derived from unsustainable practice and gain financial advantage. As discussed in the previous chapter it is clear that an alternative to the fast fashion model is required to provide a more stable platform for a sustainable fashion industry to exist on. *Slow fashion* showcases ethical and localised production, allowing retailers, particularly small independent retailers to break through into a transformational economy [54] by supporting, supplying and selling this type of production process. At this point it is important to define slow fashion so that any miss understanding does not occur. In her book *Slow fashion: aesthetics meets ethics,* [73] refers to slow fashion as sustainable design, and a business that puts people, livelihoods, and sustainability central to everything they do. The move to a slow fashion approach is not be confused with a return to six monthly cycles as the fast fashion sector of the business is too large and too complex for this to be possible. Although they cannot dramatically alter the way they work, it does not mean that a slow fashion approach cannot be implemented, there remains scope for parts of the process to *slow down*.

This slowing down can be driven by three stages:

- Transformation
- Communication
- Implementation.

A business model can be defined as "assumptions about what a company gets paid for" [29]. This means a theoretical plan based on supporting facts that are income related. It is essentially being about identifying customers and competitors, their values and behaviour and highlighting gaps that make up a company's strengths and weaknesses. This is not necessarily the case with fashion sustainability where a sense of conscious and long term planning is required. The slow business model works on small-scale production, hand-crafted techniques, social responsibility and locally sourced materials. This is often resource led or skill based is about creation and production with a fair wage paid rather price or profit driven than selling or marketing. A slow ethos contributes to transformation within the fashion industry by focusing on diversity as opposed to commercial and financial led goals. Clark [29] states that three lines of reflection are addressed; the valuing of local resources and distributed economies, transparent production systems with less intermediation between producer and consumer and sustainable and sensorial products that have a longer usable life and are more highly valued than typical consumables.

The transformation timescale to introduce a better adoption of consumer behaviour and awareness is still a medium to long-term aim. Short term projects are on-going but what is required is a more consistent critical mass of consumer opinion that will only come about if the consumer can be placed in an informed position and able to make their own decisions. The clothing and manufacturing sector needs to integrate the various models and initiatives in order to create a substantial critical mass. Current research lacks consistency in defining the route forward and in making both a transparent supply chain that is visible to the consumer with a lack of information sharing between key stakeholders such as suppliers, manufacturers, retailers and consumers.

Another area that comes indirectly under communication and that is often ignored under the larger umbrella of the fashion retailer is the fashion buyer. The person who actually runs the buying budget and makes the decisions regarding suppliers to work with, and how much product to buy in volume terms and under what conditions. The buyer's role is extremely important within retail [56] and is responsible for purchasing garments that are sold on behalf of a retailer to the consumer. Buyers have both access to the suppliers and can have contact with the consumer. They wield the power to purchase and negotiate with the suppliers and can see and influence the consumers purchasing intentions by the products they buy, promote and display on the shop floor. While most large retail department stores have in house buying teams it is the buyers that are the key in the process as they must assess manufacture to consumption accordingly [23]. By taking less of a margin they can have an enormous effect on the working conditions within the supply chain if the financial rewards are re invested. This unfortunately is not always the case (for example, Rana Plaza) and they rarely concede on margin. This in balance in buying power coinciding with the need for a more transparent supply chain accentuates the need for a new business model. The rationale for an increase in supply chain transparency is quite straight forward as it involves the proposal of a new business model where consumers are more in touch with the manufacturing supply chain [27].

Communication is becoming an increasing problem in the 21st century social media dominated arena as consumers and companies create 'alternative facts' as they fail to accept the truth or rather that they could be wrong. While the benefits of social labelling where mentioned in the previous chapter there is also research to show that even this is not straight forward [49]. The worshipping false labels mentioned in the Terra Choice report highlights that the 'sin' is on the rise for the 'worshipping false labels' where a product either through words or images gives the impression it possesses environmental attributes. This is typified by the fact that the seven sins of greenwashing still apply throughout the majority of fashion products and the consumer has little or no knowledge as to which or all of the sins has been administered. A garment that contains only one of the seven is still essentially non-compliant although is obviously a far greater achievement than one that has five areas of fault. One could argue that one is just as bad because with little effort it could eliminated although probably a particular cost in the process is the reason for its existence. The Terra Choice report goes on to mention the consumer report which categorises labels and describes their certification standards e.g., the USDA organic label. This coupled with the eco labelling website which explains what products the label is used for and the steps the producers and manufacturers must follow to obtain certification provides a solid starting point but it is one that needs to become prominent in the marketplace.

Joergens [62] when researching ethical fashion 'myth or future trend' indicated that surveys have produced evidence that consumers will reward businesses that treat their workers and the environment fairly and sanction those that do not.

However she highlighted that limited research has been conducted concerning the consumers' view on ethical issues in the fashion industry and its influence on their purchase behaviour [11, 39, 40]. It was also found that a basic lack of knowledge about the supply chain process was a fundamental reason for the type of answers to her research questionnaire. The reason for the participants' poor awareness of these issues is because they have had little media coverage. Furthermore, nobody could name either a fashion company with poor social responsibility or one with good social responsibility. Consumers could not identify a difference between brands. Where I am supposed to buy my clothes then? If I start to boycott brand X as opposed to brand Y. In the end the consumer just shifts from buying from one unethically acting company to another. Her research reinforced the message that campaigning must be more obvious like one label saying *ethically produced* and the other one *produced in a sweatshop*, thus confirming to the consumer to buy the ethically produced product. This level of information does not exist and is unlikely to as its open to legal backlash unless proven.

Leslie et al. [67] suggest that a new model of slow fashion is emerging as an alternative to globalised and fast commodity chains. This model is premised upon the slow food movement, which emphasises short production runs, local distinctiveness, quality, and sustainability. Both movements attempt to bring local factors together (including producers, designers, and consumers) and to cultivate an ethic of care.

The heart of these movements is a concept of slow living, which Parkins [84] defines as the conscious negotiation of the different temporalities which make up our everyday lives, deriving from a commitment to occupy time more attentively. It means acting in a deliberative fashion, reflecting on the impacts of one's actions on others. It is premised upon an ethic of care a concern regarding the materials that go into the products we consume, the longevity of products, and the lives of those connected to their production. Parallel to the slow food movement, designers are actively involved in educating consumers about quality and value. Another way to differentiate one's product line is to emphasise the exclusivity of each garment. Moving towards a model of slow consumption necessitates the production of new subjectivities that resist the dominance of speed [84].

6.1 Implementation of Detox Strategies

Designers have an onerous responsibility in a modified business model enabling change to a more detoxified and sustainable approach to the production and purchasing of fashion goods. Approaches such as zero waste pattern cutting, responsible fabric sourcing, designing garments with simple seams that allow for make and mend or replace panels and reducing size scales by making one size fit two, are all methods that can be incorporated into the design and production process. Further examples of other initiatives are work in progress. These include several initiatives which have been developed by large high street and sportswear brands that show a change in attitude. While to be commended, none of these actually tackle the need for a new slow fashion business model. This slowing of the industry needs to be accompanied with the detoxification of processes and practices throughout the product lifecycle. As highlighted in Sect. 3, there are many current operations that have a dramatic negative impact on both social and environmental factors. It is a more streamlined, efficient system which is needed to slow down and detox the fashion industry, resulting in a more responsible and sustainable future.

Three initiatives can be highlighted to show current levels of progress and have received favourable consumer reaction. Adidas introduced a project entitles *Sustainable Footprint* in 2012 which encouraged customers to return them for upcycling. Zara marketed part of the retail collection *Join Life*, made from sustainable materials, although this then makes one wonder how sustainable the rest of the collections are [44]. H&M in turn have through the conscious collection and by using organic fabrics shown themselves to be at the forefront of sustainable developments. Yet it can be argued that H&M are not doing enough and this is still more a marketing strategy than a need to slow down. Supporting this point Du et al. [42] recommend putting sustainability at the heart of the new product development process. The casual wear brand Patagonia was one of the first to state that 1% of annual sales from Patagonia would go to their corporate responsible movement [33].

The new business model should resist obvious routes to change consumer behaviour and awareness by additional taxes and charges but should focus on three other areas that the consumer will find easier to engage with. These being subsidies and incentives, communication and advertising and very importantly education. This is in two forms, firstly direct to Government via the all parliamentary groups, where lobbying since 2009 has made good progress. Secondly through educating young people on sustainable issues through debate, discussion workshops and seminars and introducing a new generation to sustainability and politics where they can understand how to make a difference in influencing policy change. The question still remains what progress has been made since the Rana Plaza disaster in Bangladesh in 2013.

If companies aligned themselves to a series of trade bodies and government institutions a more stable and coherent platform would exist. Global organic Textile Standards, Oeko-Tex, United Nations global compact and the Ethical trading initiative (ETI) come into this category but all with split and inconsistent membership. There is no obligation at present to do this and codes of conduct remain largely a company initiated entity. Aside from governance and policing in order to apply making sustainable practice the norm, then four straight forward areas and attitudes need to be supported to make a substantial difference.

These buy better philosophies are:

- Made in
- The setting up and managing foundation charity work to support ethical fashion communities to produce slow fashion clothing
- Crediting skills (hand- made) and developing employability
- Using and Labelling of organic and sustainable materials on garments.

In simple terms slow fashion promotes craftsmanship and quality against cheap, fast quickly made garments that are not built to last. They are essentially made to encourage the consumer to wear and repurchase within weeks. In short, it is a case of thoughtful consumption versus unthoughtful mass production. Some of the most interesting developments are from small e-commerce companies such as Cuyana, Modavanti and Zady. In a contrast of scale H&M has indicated that fast fashion brands can slow down. They have a strategy that is aimed at making consumers feel part of the brand but also allowing them to feel conscious that they are contributing to making fashion more an ethical and sustainable. Roberts [91] makes it clear that a number of high profile companies have found to their cost, corporate reputations can be significantly affected by firms management of sustainability issues, including those that are outside of their direct control, such as the environmental and social impacts of their supply networks. In a similar manner Sampson et al. [93] highlights an interesting area for research which is that of service supply chains. They point out that in traditional supply chains consumers select goods, pay for them and can provide feedback however in a new model the consumer can alter the process to their liking by taking part in much more in depth two-way feedback process that allows them to affect the decision making. It is thought that some level of customer involvement is a defining characteristic of all service organizations [66, 92] and that manufacturing firms typically have used their option of creating distance between themselves and customers to buffer core technologies and highlight that customer roles can impose costs in quality and inefficiency, which manufacturers may want to avoid [66].

To implement the new business model a new sense of trust and consistency is required. The research has shown that a series of recommendations that would come under the slow fashion category would make a significant difference to the consumer awareness and purchasing. These include:

- Buy less, the consumer is simply spoilt for choice
- Reduce the amount of 'new story drops' into retail, new product every week is simply unnecessary
- Make sure a higher percentage of goods are made from textiles that can be recycled or upcycled
- More time spent range planning by design teams to make more comprehensive decisions on the amount of styles, sizes and the colours required
- Informative packaging that can be recycled
- Greater percentages of handmade and locally sourced products.

In addition, with the new business model a company would be recommended to:

- Work with a series of trade bodies to improve production compliance
- Source organic raw materials that provide the consumer with an ethical fashion product
- Make a contribution to an ethical foundation to show the consumer a sense of sustainable conscious
- Clearly label of products based on a grading the level of ethical supply chain compliance

- Inform the consumer both at the start, middle and end of the supply chain process; e.g., at marketing, transparent making process and at retail point of sale and labelling.

The recommendations made above are based on primary and secondary research that shows the consumer must be integrated into the ethical and sustainable process through a new awareness created by a more transparent supply chain. Those companies (although mainly small SME'S) that have been brave enough to take this route are, and will be in a good position to build on the strategy to the extent that it is all they have known from day one. For those companies that are struggling to address these, it needs to should become second nature. The longer it is left it will start to seem like an inquisition or intrusion by the consumer with the danger that the consumer becomes suspicious and switches brand allegiance.

One of the key factors in the implementation of a new business model is the method to measure success. Research has found that success indicators in various fashion business models can differ wildly. Mustonen et al. [78] discuss that the objective of their study was to analyse the business models of various fashion companies, based on their 2009 financial indicators, to understand how different operation models correlate with success and whether selected performance indicators monitor operational success. They found that there was indeed substantial variation so therefore a lack of consistency. A new sense of trust would be a route to follow to start to address this key point.

7 Conclusions

The feeling of change within the fashion industry has been a prominent discussion throughout this chapter, not only a historical change and the relevant consequences but also the need for further change in the future. This sense of uncertainty has left the industry in a state of flux with a lot of both academic and external work being carried out to attempt to take the industry back to a more stable place. Overcoming challenges, of which at present there are many, will be crucial in driving the fashion industry forward. However gradual changes and development s in recent years have provided an encouraging outlook for the discipline going forward.

Consumers however have also played a vital role in this period of change and have had a great amount of influence on the direction and pace of the adaptations in the industry. The development of the fast fashion business model has triggered a change in the wants and needs of consumers, from of a preference of quality towards a more quantity based approach. This disregard for quality has also been encouraged by the growth of the value sector, with consumers now expecting to pay much less for their clothing than before. This has consequently had an impact on the value of clothing, with consumption levels soaring, garment disposal has never been higher. This poses a lot of issues in terms of environmental impacts considered at the use phase of the product lifecycle.

The changing landscape of the fashion industry has been one rationale for a lot of these changes occurring in the past decade, however the accessibility to responsible fashion goods has also been a prominent issue preventing change. The general mass-market approach to ethics and sustainability has been to add token ranges into the brand offering in preference to embedding these values into their core business practices. Change in this sense requires a more holistic and integrated approach to encourage consumer engagement in more sustainable behaviours.

It has been discussed that there are numerous ways in which the industry can strive towards a more positive future, however the consumer could also pay a big role in the change. It has been debated that the consumer does not have enough knowledge regard ethics and sustainability in order to make an informed decision. This lack of knowledge is potentially having a negative effect on the uptake of more sustainable behaviours due to a lack of knowledge and awareness. In order to address this a more transparent business approach could be taken by brands and retailers. Through the release of relevant information to their customers, companies could provide this knowledge and address this gap in awareness of responsibility in the garment supply chain. A further issue to add to this debate is the notion of applying this knowledge and awareness. It has been found that consumers who do possess knowledge regarding ethical and sustainable practices do not also implement it during their purchasing behaviour. This presents itself as a further problem currently preventing change. However it has also been found that consumer often feel a disconnect with the knowledge they possess and the garments they are purchasing. This could again be address by the retailer, through making connections between the products being sold and their core business values, consumers will begin to see the vital relationship between responsibility and production of goods.

Despite there being many problems presented during this chapter, there are also a lot of positive behaviour currently happening as well as the potential for further scope going forward. In the past price has been presented as an issues preventing further responsible behaviour change, however recent research has found that consumers are now becoming to be prepared to pay extra for a product that is manufacture responsibly. Consumer typologies have been created to help categorise people in their behaviour in the hope that a more focused approach to change can be developed. McNeil and Moore describe the ideal consumer as *sacrifice* as someone who is prepared to make change in order to adopt a more sustainable lifestyle. However it is the remaining two categories of *self* and *social* who now need to be targeted to change their priorities and ultimately their purchasing attitudes and behaviours.

Companies such as DEFRA are also addressing consumer engagement as a factor which has the potential to drive forward change. They developed the 4E's (enable, engage, encourage and exemplify) to create an adopted approach to aid further engagement between consumers and sustainable values. However it is consumers and companies alike that need to drive forward for change in order to create an impact.

A slow approach has been suggested with more focus being on the value created within the garment through considerations such as small scale production methods, hand crafted elements and locally sourced materials. This move away from a constant purchasing of cheap, low quality garments in preference for more considered purchasing in the hope of improved product longevity and the creation of enhanced care and value.

References

1. Accord (2014) Accord on fire and building safety in Bangladesh. Available at: www.bangladeshaccord.org. Accessed 30 Jan 2014
2. Ahmed F (2012) At least 117 killed in fire at Bangladeshi clothing factory. CNN, 25 Nov. Available at: http://edition.cnn.com/2012/11/25/world/asia/bangladesh-factory-fire/?hpt=hp_t1. Accessed 30 Jan 2013
3. Ajzen I (1985) From intentions to actions: a theory of planned behaviour. Springer-Verlag, New York
4. Alam J (2012) 112 killed in fire at Bangladesh garment factory. The Associated Press, 25 Nov 2012. Available at: http://bigstory.ap.org/article/112-killed-fire-bangladesh-garment-factory-0. Accessed 30 Jan 2014
5. Ali Manik J, Yardley J (2013) Building collapse in Bangladesh leaves scores dead. The New York Times, 24 Apr 2013. Available at: http://www.nytimes.com/2013/04/25/world/asia/bangladesh-building-collapse.html?hp&_r=1&. Accessed 24 Apr 2013
6. Arnold C (2009) Ethical marketing and the new consumer. Wiley, West Sussex
7. Asia Floor Wage (2009) Asia floor wage - stitching a decent wage across borders. Available at: www.asiafloorwage.wordpress.com/about/. Accessed 30 Jan 2014
8. Auger P, Devinney TM (2007) Do what consumers say matter? the management of preferences with unconstrained ethical intentions. J Bus Ethics 76:361–383
9. Baggini J, Fosl P (2007) The ethics toolkit. Wiley—Blackwell, Boston
10. Bajaj V (2012) Fatal fire in Bangladesh highlights the dangers facing garment workers. The New York Times, 25 Nov 2012. Available at: http://www.nytimes.com/2012/11/26/world/asia/bangladesh-fire-kills-more-than-100-and-injures-many.html?_r=0. Accessed 30 Jan 2014
11. Balzer M (2000) Gerechte Kleidung: fashion Oko fair. Hirzel Verlag, Stuttgart
12. Barrie L (2009) Just style. Available at: www.just-style.com/comment/spotlight-onsustainability-in-design_id105739.aspx. Accessed 11 Jan 2011
13. BBC (2013a) Bangladesh building collapse death toll over 800. BBC News, 8 May 2013. Available at: http://www.bbc.co.uk/news/world-asia-22450419. Accessed 8 May 2013
14. BBC (2013b) Bangladesh building collapse death toll passes 500. BBC News, 3 May 2013. Available at: http://www.bbc.co.uk/news/world-asia-22394094. Accessed 3 May 2013
15. BBC (2012) Dhaka Bangladesh clothes factory fire kills more than 100. BBC News, 25 Nov 2012. Available at: http://www.bbc.co.uk/news/world-asia-20482273. Accessed 30 Jan 2013
16. Belk R, Devinney TM, Eckhardt G (2005) Consumer ethics across cutures. Consumption, Mark Cult 8:275–289
17. Bergman D, Blair D (2013) Bangladesh: Rana Plaza architect says building was never meant for factories. The Telegraph, 3 May 2013. Available at: http://www.telegraph.co.uk/news/worldnews/asia/bangladesh/10036546/Bangladesh-Rana-Plaza-architect-says-building-was-never-meant-for-factories.html. Accessed 8 May 2013

18. Bhuiyan K (2012) Bangladesh garment factory disaster timeline. Compliance - Updates, 29 November 2013 [Online]. Available at: www.Steinandpartners.com/sustainability/compliance/bangladesh-garment-factorydisaster-timeline. Accessed 30 Jan 2014

19. Bray J, Johns N, Kilburn D (2010) An exploratory study into the factors impeding ethical consumption. J Bus Ethics 98:597–608

20. British Fashion Council (2013) About Estethica. Available at: www.britishfashioncouncil.com/content/1146/estethica. Accessed 14 Apr 2013

21. Brosdahl JC (2007) The consumption crisis. In: Hoffman L (ed) Future fashion. The White Papers, Earth Pledge, pp 46–57

22. Butler S (2013) Bangladeshi factory deaths spark action among high-street clothing chains. The Guardian, 23 June 2013. Available at: http://www.theguardian.com/world/2013/jun/23/rana-plaza-factory-disaster-bangladesh-primark. Accessed 26 Aug 2013

23. Callon M, Meadel C, Rabehari V (2002) The economy of quantities economy and society. Econ Soc 31(2):194–217

24. Carey L, Shaw D, Shiu E (2008) The impact of ethical concerns of family consumer decision-making. Int J Consum Stud 32:553–560

25. Carrigan M, Attalla A (2001) The myth of the ethical consumer–do ethics matter in purchase behaviour? J Consum Mark 18:560–578

26. Carrington M, Neville B Whitwell G (2010) Why ethical consumers dont walk their talk: towards a framework for understanding the gap between the ethical purchase intentions and actual buying behaviour of the ethically minded consumers. J Bus Ethics 97(1):139–158

27. Castaldi C, Dickson M, Grover C (2013) Sustainability in fashion and textiles, values. Greenleaf publishing, Sheffield

28. Chatzidakis A, Hibbert S, Smith AP (2007) Why people dont take their concerns about fair trade to the supermarket: the role of neutralisation. J Bus Ethics 74(1):89–100

29. Clark H (2008) Slow + Fashion—an oxymoron—or a promise for the future …? fashion theory. J Dress Body Cult 12(4):427–446

30. Clean Clothes Campaign (2012) One year after Tazreen fire, the fight for justice continues. Available at: https://www.cleanclothes.org/news/press-releases/2013/11/21/one-year-after-tazreen-fire-the-fight-for-justice-continues. Accessed 30 Jan 2013

31. Clouder S, Harrison R (2005) The effectiveness of ethical consumer behaviour. In: Harrison R, Newholm T, Shaw D (eds) The ethical consumer. Sage Publications, London, pp 89–106

32. Cowe R, Williams S (2001) Who are the ethical consumers?. Co-operative Bank, Manchester

33. Croll J (2014) Fashion that changed the world. Prestel, New York

34. Dahl R (2010) Green washing do you know what you are buying? Environ Health Perspect 118(6):246–252

35. Davies C (2007) Branding the ethical consumer. The Financial Times, 21 Feb, p 18

36. Devinney TM, Auger P, Eckhardt G (2010) The myth of the ethical consumer. Cambridge University Press, Cambridge

37. Devinney T, Auger P, Eckhardt GM (2011) 'Values vs. Value', strategy and business, 22 Feb 2011. Available at: http://www.strategy-business.com/article/11103?gko=03d29. Accessed 12 Jan 2017

38. Devnath A, Srivastava M (2013) Suddenly the floor wasn't there. Factory survivor says', Bloomberg, 25 Apr 2013. Available at: http://www.bloomberg.com/news/2013-04-25/-suddenly-the-floor-wasn-t-there-factory-survivor-says.html. Accessed 28 Apr 2013

39. Dickson M (1999) US knowledge of concern with apparel sweatshops. J Fashion Mark Manag 3(1):44–55

40. Dickson M (2000) Personal values, beliefs, knowledge and attitudes relating to intentions to purchase apparel from socially responsible businesses. Clothing Text Res J 18(1):19–30

41. Dickson MA (2013) Identifying and understanding ethical consumer behavior: reflections on 15 years of research. In: Bair J, Dickson M, Miller D (eds) Workers' rights and labor compliance in global supply chains. New York Routledge, pp 121–139

42. Du S, Yalcinkaya G, Ludwig B (2016) Sustainability, social media driven open innovation, and new product development performance. J Prod Innov Manag 33:55–71
43. Ellen P (1994) Do we know what we need to know? objective and subjective knowledge effects on pro-ecological behaviours. J Bus Res 30(1):43–52
44. Ethlers S (2016) Zara join life campaign. Fashion United, 20 Sept 2016. Available at: https://fashionunited.uk/news/fashion/zara-goes-sustainable-with-new-join-life-initiative/2016092021831. Accessed 16 Jan 2017
45. Fashioning an Ethical Industry (2010) Fashioning an ethical industry. Available at: www.fashioninganethicalindustry.eu. Accessed 13 Apr 2011
46. Fletcher K (2008) Sustainable fashion and textiles. Earthscan, London
47. Fletcher K, Tham T (2014) The handbook of sustainability and fashion. Routledge, London
48. Fletcher K (2010) Slow fashion. An invitation for systems change. J Des Prac Proc Fashion Ind 2(2):259–265
49. Fliegelman J (2010) The next generation of greenwash: diminishing consumer confusion through a national eco-labelling program. Urban Law J 37(4):1001
50. Fox K (2005) Watching the English. Hodder & Stoughton, London
51. Foxall G (2005) Understanding consumer choice. Palgrave Macmillian, Hampshire
52. Freestone O, Mcgoldrick P (2007) Motivations of the ethical consumer. J Bus Ethics 79(4):445–467
53. Gardetti M, Torres A (2015) Sustainable luxury, managing social and environmental performance in iconic brands. Greenleaf publishing, Sheffield
54. Gardien P, Djajadiningrat T, Hummels C, Brombacher A (2014) Changing the hammer: the implications of paradigmatic innovation of design practice. Int J Des 8(2):119–139
55. Genecon LLP, Partners (2011) Understanding high street performance. 5 Dec. Available at: www.gov.uk/government/uploads/system/uploads/attachment_data/file/31823/11-1402-understanding-high-street-performance.pdf. Accessed 12 Apr 2012
56. Goworek H (2007) Fashion buying. Blackwell, London
57. Harney A (2008) The China price—the true cost of Chinese competitive advantage. The Penguin Press, New York
58. Hawkes S (2013) People thought Rana Plaza would be a blow to Primark. Today's profit figures say otherwise. The Telegraph, 5 Nov 2013. Available at: http://blogs.telegraph.co.uk/news/stevehawkes/100244468/people-thought-rana-plaza-would-be-a-blow-to-primark-todays-profit-figures-say-otherwise/. Accessed 30 Jan 2013
59. Hawkins DE (2006) Corporate social responsibility. Palgrave MacMillan, Hampshire
60. Jackson T (2006) The earthscan reader in sustainable consumption. Earthscan, London
61. James A, Montgomery B (2016) The role of the retailer in socially responsible fashion purchasing. In: Muthu SS (ed) Textiles and clothing sustainability: sustainable fashion and consumption. Springer, Hong Kong
62. Joergens C (2006) Ethical fashion myth or future trend. J Fashion Mark Manage 10(3):360–371
63. Kim A, Ko E (2012) Do social media marketing activities enhance customer equity? an empirical study of luxury fashion brand. J Bus Res 65(10):1480–1486
64. Lee M (2007) Eco Chic. Octopus Publishing Group, London
65. Lee M, Sevier L (2007) Green pages—ethical fashion special. The Ecologist, p 66
66. Lengnick-Hall C (1996) Customer contributions to quality: a different view of the customer oriented firm. Acad Manag Rev 21(3):791–824
67. Leslie D, Brail S, Hunt M (2014) Crafting an antidote to fast fashion: the case of Toronto's independent fashion design sector. Growth Change 45(2):222–239
68. McCartney S (2016) Q&A. Available at http://www.stellamccartney.com/experience/en/sustainability/qa-with-stella/. Accessed 23 Jan 2017
69. Mcmullen A, Maher S (2009) Lets clean up fashion 2009—the state of pay behind the UK high street. Labour Behind the Label, Bristol
70. Miller D (2013) Last nightshift in Savar: the story of the spectrum sweater factory collapse. McNidder and Grace Limited, Alnwick

71. Miller D (2010) Social labelling in the global fashion industry. Available at: http://socialalterations.com/wpcontent/uploads/2009/11/Social_Labelling_in_the_Global_Fashion_Industry1.pdf. Accessed 25 Jan 2017

72. Ming L, Zhang Z, Leung C (2004) Fashion change and fashion consumption. J Fashion Mark Manage 8:362–374

73. Minney S (2016) Slow fashion: aesthetics meets ethics. New Internationalist Publications ltd, Oxford

74. Mintel (2007) Green and ethical consumer. Mintel, London

75. Mintel (2009) Ethical clothing. Available at: http://academic.mintel.com. Accessed 27 Nov 2010

76. Mol AP (2015) Transparency and value chain sustainability. J Clean Prod 107:154–161

77. Morgan L, Birtwhistle G (2009) An investigation of young consumers disposal habits. Int J of Consum Stud 33(2):190–198

78. Mustonen M, Palb R, Mattila H, Mashkoor Y (2013) Success indicators in various fashion business models. J Glob Fashion Mark 01:1–19

79. Nelson D, Bergman D (2013) Scores die as factory for clothing stores collapses. The Independent 25 Apr 2013. Available at: http://www.independent.ie/world-news/asia-pacific/scores-die-as-factory-for-clothing-stores-collapses-29220894.html. Accessed 25 Apr 2013

80. Newholm T, Shaw D (2007) Studying the ethical consumer: a review of research. J Consum Behav 6:253–270

81. Niinimaki K (2010) Eco-clothing, consumer identity and ideology. Sustain Dev 18:150–162

82. Ozcaglar-Toulouse N, Shiu E, Shaw D (2006) In search of fair trade: ethical consumer decision making in France. Int J Consum Stud 30:502–514

83. Papaoikonomou E, Ryan G, Ginieis M (2011) Towards a holistic approach of the attitude behaviour gap in ethical consumer behaviours: empirical study from Spain. Int Adv Econ Res 17(1):77–88

84. Parkins W (2004) Out of time: fast subjects and slow living. Time Soc 13(2/3):363–382

85. Paul R, Quadir S (2013) Bangladesh urges no harsh EU measures over factory deaths. 4 May 2013. Available at: http://www.reuters.com/article/2013/05/04/us-bangladesh-factory-idUSBRE94304420130504. Accessed 30 Jan 2014

86. Portas M (2011) The portas review: an independent review into the future of our high streets. Available at: http://www.bis.gov.uk/assets/biscore/business-sectors/docs/p/11-1434-portas-review-future-of-high-streets.pdf. Accessed 4 Feb 2012

87. Preez R (2003) Apparel shopping behaviour—part 1: towards the development of a conceptual theoretical model. J Ind Psychol. Available at: http://www.sajip.co.za/index.php/sajip/article/view/111/107. Accessed 10 Oct 2010

88. Reilly AH, Hynan KA (2014) Corporate communication, sustainability, and social media: it's not easy (really) being green. J Bus Horiz 57(6):747–758

89. Rest JR (1986) Moral development: advances in research and theory. Praeger, New York

90. Ritch E, Schroder M (2009) Whats in fashion? ethics? an exploration of ethical fashion consumption as part of modern family life. Available at: http://www.northumbria.ac.uk/static/5007/despdf/events/era.pdf. Accessed 24 Sep 2010

91. Roberts S (2003) Supply chain specific?, understanding the patchy success of ethical sourcing initiatives. J Bus Ethics 12:159

92. Sampson S, Froehle C (2006) Foundations and implications of a proposed unified services theory. Prod Oper Manage 15(2):329–343

93. Sampson S, Spring M (2012) Customer roles in service supply chains and opportunities for innovation. J Supply Chain Manage 48(4):30–50

94. Schiffman L, Kanuk L, Hansen H (2008) Consumer behaviour—a European outlook. Pearson Education, Essex

95. Shields R (2008) The last word on disposable fashion. The Independent on Sunday, 28th December, p 12

96. Singer P (1997) How are we to Live? ethics in the age of self interest. Oxford University Press, Oxford

97. Smith NC (1995) Maketing strategies for the ethics era. Sloan Manage Rev 36:85–98
98. Soloman M, Rabolt N (2004) Consumer behaviour in fashion. Pearson Education, New Jersey
99. Strong C (1996) Features contributing to the growth of ethical consumerism—a preliminary investigation. Market intelligence and planning. Available at: http://www.emeraldinsight.com/journals.htm?articleid=854342&show=abstract. Accessed 1 Oct 2010
100. Szmigin I, Carrigan M, Mceachern G (2009) The conscious consumer: taking a flexible approach to ethical behaviour. Int J Consum Stud 33(2):224–231
101. TEDxSalford (2014) The wardrobe to die for—Lucy Siegle. Available at: https://www.youtube.com/watch?v=YglyHzvBqpA. Accessed 26 Jan 2017
102. The True Cost (2015) The true cost. Available at: http://truecostmovie.com. Accessed 26 Jan 2017
103. Tokatli N (2007) Global sourcing: insights from the global clothing industry—the case of Zara, a fast fashion retailer. Journal of Economic Geography. Available at: http://www.nihul.biu.ac.il/_Uploads/dbsAttachedFiles/zara06.pdf. Accessed 30 Sept 2010
104. United Nations—World commission on environment and development (1987) Report of the world commission on environment and development: our common future. Available at: http://www.un-documents.net/our-common-future.pdf. Accessed 26 Jan 2017
105. Wehrmeyer W (1992) Strategic issues. In: Charter M (ed) Greener marketing. Sheffield Greenleaf, pp 41–56
106. Wgsn (2010) Final edit 2010. World Global Sourcing Network, 28 Sept. Available at: www.wgsn.com. Accessed 29 Nov 2011
107. Worcester R, Dawkins J (2005) Surveying ethical & environmental attitudes. In: Harrison R, Newholm T, Shaw D (eds) The ethical consumer. Sage Publications, London, pp 189–203
108. WWF (2015) The hidden cost of water. WWF. Available at: http://www.wwf.org.uk/what_we_do/rivers_and_lakes/the_hidden_cost_of_water.cfm. Accessed 1 Mar 2016
109. Zain Al-Mahmood S, Smithers R (2013) Matalan supplier among manufacturers in Bangladesh building collapse. The Guardian, 24 Apr 2013. Available at: http://www.theguardian.com/world/2013/apr/24/bangladesh-building-collapse-kills-garment-workers. Accessed 24 Apr 2013

Erratum to: Detox Fashion

Subramanian Senthilkannan Muthu

Erratum to:
S.S. Muthu (ed.), *Detox Fashion*,
Textile Science and Clothing Technology,
https://doi.org/10.1007/978-981-10-4777-0

In the original version of the book, the Chapters "Detoxifying Luxury and Fashion Industry: Case of Market Driving Brands" and "Integrating Sustainable Strategies in Fashion Design by Detox 2020 Plan—Case Studies from Different Brands" have been replaced with new Chapters "Environmental Issues in Textiles: Global Regulations, Restrictions and Research" and "Making the Change: The Consumer Adoption of Sustainable Fashion".

The updated online version of this book can be found at
https://doi.org/10.1007/978-981-10-4777-0_2
https://doi.org/10.1007/978-981-10-4777-0_3
https://doi.org/10.1007/978-981-10-4777-0

Printed in the United States
By Bookmasters